RISK, CHOICE, AND UNCERTAINTY

RISK, CHOICE, AND UNCERTAINTY

THREE CENTURIES OF ECONOMIC DECISION-MAKING

George Szpiro

COLUMBIA UNIVERSITY PRESS

NEW YORK

Columbia University Press
Publishers Since 1893
New York   Chichester, West Sussex
cup.columbia.edu
Copyright © 2020 Columbia University Press
All rights reserved

Library of Congress Cataloging-in-Publication Data
Names: Szpiro, George, 1950– author.
Title: Risk, choice, and uncertainty : three centuries of economic decision-making / George G. Szpiro.
Description: New York : Columbia University Press, 2019. |
Series: Risky decisions | Includes bibliographical references and index.
Identifiers: LCCN 2019023645 (print) | LCCN 2019023646 (e-book) |
ISBN 9780231194747 (cloth) | ISBN 9780231550970 (e-book)
Subjects: LCSH: Decision making. | Risk management.
Classification: LCC HD30.23 .S97 2019 (print) | LCC HD30.23 (e-book) |
DDC 330.01/9—dc23
LC record available at https://lccn.loc.gov/2019023645
LC e-book record available at https://lccn.loc.gov/2019023646

Columbia University Press books are printed on permanent and durable acid-free paper.
Printed in the United States of America

Cover design: Noah Arlow
Cover image: ©Jorg Greuel/Getty Images

*This book is dedicated to Tamar, Rotem and Daniel
and to all who will follow.*

CONTENTS

Introduction ix

PART I: HAPPINESS AND THE UTILITY OF WEALTH

1. IT ALL BEGAN WITH A PARADOX 3
2. MORE IS BETTER . . . 23
3. . . . AT A DECREASING RATE 41

PART II: MATHEMATICS IS THE QUEEN OF THE SCIENCES . . .

4. THE MARGINALIST TRIUMVIRATE 67
5. FORGOTTEN PRECURSORS 102
6. BETTING ON ONE'S BELIEF 109
7. GAMES ECONOMISTS PLAY 119
8. WOBBLY CURVES 138
9. COMPARING THE INCOMPARABLE 152

PART III: ... BUT MAN IS THE MEASURE OF ALL THINGS

10. MORE PARADOXES   165

11. GOOD ENOUGH   179

12. SUNK COSTS, THE GAMBLER'S FALLACY, AND OTHER ERRORS   188

13. ERRONEOUS, IRRATIONAL, OR PLAIN DUMB?   206

Notes   215
Bibliography   229
Index   239

INTRODUCTION

Two schools of Jewish thought prevailed in the first century BCE, led by Rabbi Hillel and Rabbi Shammai, respectively. The Talmud tells that the latter was strict and impatient and had a short temper, while Hillel was gentle and conciliatory. One day, a stranger came to Shammai and asked to be converted to Judaism. He was in a great hurry and demanded: "Make me a Jew on condition that you teach me the whole Torah while I stand on one foot." What nerve! Not yet Jewish, and already showing such chutzpah! Incensed, Shammai hit the man with a cane and threw him out of his house. Unperturbed, the stranger went to Hillel to try his luck with him. Calmly, Hillel took up the challenge. While the insolent would-be Jew stood on one foot, the rabbi declared: "What is hateful to you, do not do unto others. That is the whole Torah; all the rest is commentary. Now go and study."

Why do I tell this anecdote? Well, economics is similar: People prefer more of a good, but the more they already have, the less they value each additional unit. That's the whole of economics; all the rest is commentary. Now go and read.

The fact that people prefer to possess more of nearly everything, but that they do so at a decreasing rate, already had been observed by the ancient Greeks. Much later, in the eighteenth century, the mathematician Daniel Bernoulli, then teaching in St. Petersburg in Russia, used this knowledge to propose a solution to a gambling problem posed by his cousin Nikolaus. The ingenious solution that he and—independently—his fellow mathematician Gabriel Cramer proposed was based precisely on the fact that the utility of wealth (i.e., its enjoyment) diminishes as wealth increases. After all, an additional dollar offers more utility to a homeless beggar than to a millionaire.

For these mathematicians, the St. Petersburg Paradox represented nothing more than an interesting problem about games of chance; they drew no connection between the mathematical conundrum and economics. Indeed, eighteenth-century economics was mostly about casual observations. Adam Smith, for example, who laid the foundations of classical economics, introduced the notions of division of labor and economies of scale by describing work in a needle factory. Early economists did suggest models of the economy. However, apart from arithmetic illustrations and examples, their work consisted mainly of words: They described their observations, recounted anecdotes, and explained their conclusions.

Thus, compared to, say, physics, medicine, or chemistry, economics was not considered a serious science—not until mathematics entered the scene, anyway. The discipline turned into a serious science only after mathematical models were developed that suggested how to optimize something, be it wealth, profits, or the utility for money. This occurred only in the late nineteenth century, when neoclassical economists began to utilize mathematical methodology and tools. (When Alfred Nobel established the Nobel prizes in 1895, economics was not among the sciences honored.)

Here is where the Bernoulli family comes in once again. Apart from providing a solution to the St. Petersburg Paradox, Daniel Bernoulli and several other members of his illustrious family of scientists were pioneers in the application of infinitesimal calculus, the mathematical techniques that had been developed by Isaac Newton and Gottfried Wilhelm Leibniz to study the continuous change of physical objects (i.e., the paths of stars and moving objects). In the late nineteenth century, three economists working independently of each other—Stanley Jevons in England, Léon Walras in Switzerland, and Carl Menger in Austria—began to apply calculus to economic science. With this, the mathematization of economics began, and a very fruitful period for the discipline ensued. But over time, the mathematics that were employed became more and more sophisticated, and eventually academic papers in economics were barely distinguishable from research in the mathematical sciences. Economics had nearly turned into a branch of pure mathematics.

This changed again in the late twentieth century, with the emergence of behavioral economics. After about the 1970s, the discipline adopted a much more behavior-oriented approach. Thus, during the past half-century or so, the aim has become to describe how people actually behave, with the discipline relying more on psychology and less on mathematics. (To wit: in the early decades after the institution of the Nobel Prize in Economics in 1968, the prize was awarded only for mathematical theories; but lately, it has been fashionable to award it for nonmathematical models.)

This book is about how people make decisions. And because decision-making underlies most, if not all, of economics, the book presents a novel approach to the history of economic thought.

It is divided into three parts. Part I sets the stage by introducing the eighteenth- and early nineteenth-century characters and their theories of utility for wealth. Part II presents the people that followed and the models they developed, mainly in the late nineteenth and first part of the twentieth century. These models guide rational actors to the *best* decisions (i.e., normative economics). Mathematics and axiomatic systems were de rigeur. Part III, which covers the developments and the personalities of the latter part of the twentieth century until today, describes how human beings actually *do* act (i.e., positive economics). It turns out that, in general, people are not rational. Hence, the beautiful mathematical models are largely ignored, and psychology is the name of the game.

The viewpoint of this book is not that economics has become too mathematically oriented and that a backlash is occurring. On the contrary—mathematics is, and must remain, the basis for normative economics—that is, how decisions *should* be made. And the behavioral models, in fact, also rely on mathematics.

However, when describing how people actually make decisions, mathematics becomes less prominent. According to a recent Nobel laureate in economics, Richard Thaler, as well as others, it is precisely the limits to decision-makers' computational abilities and their ignorance of the normative models that make humans seem "irrational."

I want to thank Anna-Maria Sigmund for transcribing some difficult-to-read, handwritten letters of Oskar Morgenstern. Thanks also go to Myles Thompson from Columbia University Press for the encouragement he afforded me, and Brian Smith, also from Columbia University Press, and Ben Kolstad from Cenveo, for their painstaking editing. I express my gratitude to the Bogliasco Foundation in New York: a residency at the foundation's beautiful estate near Genova, on the Mediterranean shore, afforded me the leisure to gather my thoughts and work on the manuscript.

As in all my undertakings, uxorial thanks go to Fortunée, who claims that she cannot do math but usually gets her calculations right . . . within a plus or minus 10 percent error band. *Life would be no fun without you!*

It is the hope of this author that this book will not only inform and entertain its readers but also make them aware of their biases, cognitive and otherwise . . . which are ultimately what makes us all human.

RISK, CHOICE, AND UNCERTAINTY

PART ONE

HAPPINESS AND THE UTILITY OF WEALTH

CHAPTER 1

IT ALL BEGAN WITH A PARADOX

On September 9, 1713, the Swiss mathematician Nikolaus Bernoulli took a seat at his desk to write a letter to an acquaintance in France, the nobleman Pierre Rémond de Montmort. A question posed therein—formulated as an innocent mathematical puzzle—would spawn an important (if not the most important) concept of human decision-making, thus setting in motion the development of economics as a science.

A scion of the renowned Bernoulli family in Basel, Switzerland, which, over three generations, brought forth no less than eight of the most celebrated men of science of all time, Nikolaus was destined to follow in his relatives' footsteps. His father's two brothers, Johann and Jakob; his cousins, Nikolaus II, Johann II, and Daniel (figure 1.1); and his cousins once removed, Jakob II, Daniel II, and Johann III, were instrumental in developing seventeenth- and eighteenth-century physics and mathematics, particularly in the field of differential calculus. Only Isaac Newton in Cambridge, Gottfried Wilhelm Leibniz in Hanover, and the Bernoullis' close friend and disciple, fellow Swiss mathematician Leonhard Euler in Basel, could measure up to the family's accomplishments.

Because there were not enough university posts for all the Bernoullis, they had to apply their talents to all kinds of intellectual pursuits. Nikolaus Bernoulli, for example, became *doctor juris* at the University of Basel with a dissertation on the application of probability theory to questions arising in the practice of law. His *De Usu Artis Conjectandi in Jure* (On the use of the

FIGURE 1.1: **Daniel Bernoulli.**

Source: Wikimedia Commons

technique of conjecturing in matters of law) dealt not only with statistical intricacies of annuities, legacies, lotteries, insurance, and the credibility of witnesses but also with questions such as when a missing person, given his life expectancy, can be considered dead. The book's last remark is especially noteworthy: "Counselors at law are [even] able to argue over goat's wool." In spite of—or perhaps because of—such sharp insights, he was appointed professor of law in 1731. Regrettably, the family's history is not without its conflicts.

Intense competition, extreme rivalry, and petty priority disputes occasionally pitted brother against brother, father against son.[1]

Nikolaus began his letter to Montmort by apologizing for not having written earlier: "For quite some time, I have been unable to do any new research about the subject matter of chance, which is why I have nothing to report to you. Nevertheless, in compensation for the problems that you suggested to me and whose solutions I will examine as soon as I have free time, I propose some others that deserve your attention."[2] He then posed five questions about the chances of winning at a game of dice, the last two of which are relevant to our story:

> Fourth Problem: *A* promises to give *B* one coin if he throws six points at his first throw, two coins if he gets six at the second throw, three if he gets it at the third throw, and so on. The question is "What is *B*'s expectation?"
>
> Fifth Problem: The same question, except that this time, instead of giving 1, 2, 3, 4 . . . coins, as before, *A* promises to give *B* the amounts 1, 2, 4, 8, 16 . . . or 1, 3, 9, 27 . . . or 1, 8, 27, 64 . . .

"Even though most of these problems are not difficult, you will find something very interesting in them," Nikolaus wrote, and ended his letter with the flowery verboseness that was common at the time: *"J'ai l'honneur d'être avec un attachement inviolable, MONSIEUR, Votre très humble & très obéissant Serviteur N. Bernoully."*[3]

By asking *"quelle est l'espérance de B?"* (What is B's expectation?) in the fourth and fifth problems, Nikolaus referred to the notion of *mathematical expectation*—a novel notion at the time. The concept had been coined only about sixty years earlier, near the middle of the seventeenth century, in a correspondence between two amateur mathematicians (albeit amateurs of the first order): the novelist Blaise Pascal in Paris, and Judge Pierre de Fermat in Toulouse, in southern France.[4] The exchange of letters was prompted by a query from a notorious gambler, Antoine Gombeaud, who had adopted for himself the title "Chevalier de Méré." The faux chevalier wanted to know how the content of a pot should be distributed among two players if a game of chance that should have been played for several rounds were stopped before any of the players had accrued sufficient points. For example, let us say that Blaise and Pierre are playing a game of cards for five rounds and whoever wins three rounds gets the pot. After three rounds, with the score two for Blaise and one for Pierre, both players are too drunk to continue. The game is terminated and the question now is: How should the pot be divided? Half-and-half, because each of the players could still win? Three shares to two, according to the rounds played thus far? Or two to one, as Blaise needs only one more win while Pierre needs two?

The correct answer is "None of the above." But this is far from obvious, and it was especially obscure in the 1650s, when the question was first broached. In the intensive correspondence that ensued, Pascal and Fermat expended enormous mental effort and eventually came up with the correct answer. After all, predicting what the future holds, considering the possible outcomes of unplayed games, had never been attempted before.

One could assume the following possibilities:

➤ If the game continues and the fourth round goes to Blaise, he has the necessary third point and wins the pot. (Possibility A)
➤ If it goes to Pierre, the score is now tied, and a fifth round must be played. If Blaise wins it, he now has the third point and wins the pot. (Possibility B)
➤ However, if he loses it, then it is Pierre who has the necessary three points and gets the pot. (Possibility C)

So it would seem that there are two chances out of three (A and B) that Blaise would win the game, and one chance (C) that Pierre would win it. The pot should be divided accordingly. Correct? Well, no! The elusive fact, which Pascal and Fermat found difficult to grasp at first, is that there are not three possible outcomes (A, B, and C) but four. Because the game was scheduled to go on for five rounds, it is not over, strictly speaking, even if Blaise won the fourth round. The fifth round should be played, no matter what. It could go either to Blaise (Possibility A1) or to Pierre (Possibility A2). Both cases would give Blaise a win. Hence, he has three chances of winning (A1, A2, and B), while Pierre has only one (C). So that means that 75 percent of the pot should go to Blaise and 25 percent to Pierre. The important point to note is that in order to determine Blaise's and Pierre's chances of winning, *all* possible outcomes must be taken into account, even the round that did not need to be played because Blaise already had the necessary third point.

It was a revolutionary idea. For the first time, possible outcomes of future events were taken into account, and their chances computed. It was the dawn of probability theory.

What Pascal and Fermat had come up with in their correspondence was that whenever the game is stopped, the pot should be divided according to the *expectations* that the two players have at that time. Let us say the pot contains 100 ducats. We have just seen that after the third round, and without playing the fourth and fifth ones, Blaise should be paid 75 ducats, and Pierre 25 ducats. These sums represent the amounts that Blaise and Pierre *expect* to win, in spite of the fact that if the game had been played until the end, one of them would have received 100 ducats and the other nothing. If the game were played many, many times, and the players interrupted it for a moment of reflection every time the score stood at two to one, and then continued, the player with two points would, on average, win 75 ducats, and the other would get 25 ducats. More generally, the *mathematical expectation* of a game, each time it is played, is the winner's payout (100 ducats), multiplied by the probability of a win (75 percent).[5]

It is to this notion that Nikolaus referred in his letter when he mentioned *l'espérance*. Let us say that you participate in a game of chance in which there is a 20 percent probability of winning $100, and nothing otherwise. If this game is played over and over again, 80 percent of the time you will receive nothing. In the remaining 20 percent, the payout is $100. Thus, on average, you receive $20 (i.e., 20 percent of $100). This is what is meant by the expectation: the win that you expect every time, just before the game is actually played . . . even though in real life, you will end up empty-handed eight out of ten times.

OK—back to Nikolaus Bernoulli's letter. The addressee, Rémond de Montmort, was also a mathematician of repute, if not of the stature of the Bernoullis. Five years earlier, he had published a well-regarded book, *Essay d'analyse sur les jeux de hazard* (Essay on the analysis of games of chance), in which he considered combinatorial problems as they appear in card and dice games. In an era when most of his contemporaries thought that any forecast about the outcome of a game not yet played was ludicrous and the prediction of future events the realm of witchcraft and astrology, Montmort's book was a trailblazer. The chance that something would happen was not even defined—hence, to claim that probabilities of future events, a concept as new as it was controversial, could be computed precisely with the tools of mathematics was unorthodox, if not downright blasphemous. Against this background, Montmort's *Analyse* was one of the early treatises on the theory of probability.

Montmort knew exactly what Nikolaus meant by expectation, and his response followed two months later, on November 15, 1713: "The last two of your five problems present no difficulty whatsoever. All one has to do is find the sums of the series whose numerators are progressions of squares, cubes, etc., and whose denominators follow a geometric progression. Your late uncle introduced the method to find the sum of such series." The ten-page letter ends in no less a flowery way than Bernoulli's, except that the ironic nobleman decided to add a tongue-in-cheek remark: "Some authors never finish and always have a thousand things to say that they are convinced are very useful but that one could very easily do without. In order not to risk falling into this trap, I end by assuring you that I have the greatest respect for you and am, with all my heart, Monsieur, your very humble and obedient servant, R. D. M."

The very humble and obedient servant should have taken the time to perform the calculations instead of casually dismissing the problems. "It is true, as you say, that the last two of my problems present no difficulty," Nikolaus responded on February 20, 1714, but then he added reproachfully, "Nevertheless you would have done well to find the solution because it would have furnished you the occasion to make a curious observation." He then proceeded to show something that, to a mathematician, was truly amazing. Reframing the problem from throwing a die to the simpler case of tossing a coin, I will demonstrate what it was that Bernoulli found so curious.

Let us say that Peter offers to pay Paul $1 if a coin lands heads on the first toss. If the first is tails, and the second is heads, Paul gets $2. If the first two tosses are tails, and heads appears only on the third toss, Paul will get $4. If there are three tails in a row, and only then does the coin come up heads, the payout will be $8, and so on. That is, the payout doubles each time that heads does not appear in consecutive tosses. So, what can one expect to win in such a game?

As Pascal and Fermat demonstrated, the expected win is calculated as follows: The chances of the coin landing heads on the first throw is one in two. The probability of the coin landing tails on the first toss and heads only on the second is one in four, the probability that heads will appear only on the third toss, after two tails, is one in eight, and so on. If, say, the coin lands on tails ten times in a row, and heads appears on the eleventh throw, the payout would be a cool $1,024. However, the probability of such a series of throws—ten times tails, followed by heads—is very low. On average, such an event is expected to happen only once every 2,048 games.

The expected win is the sum of the individual payouts (1, 2, 4, 8, 16 . . .) multiplied by the probabilities ($\frac{1}{2}$, $\frac{1}{4}$, $\frac{1}{8}$, $\frac{1}{16}$, $\frac{1}{32}$, . . .). Hence, we have

$$\begin{aligned} \text{Expected win} &= (1 \times \tfrac{1}{2}) + (2 \times \tfrac{1}{4}) + (4 \times \tfrac{1}{8}) + (8 \times \tfrac{1}{16}) + (16 \times \tfrac{1}{32}) + \ldots \\ &\quad + (1{,}024 \times \tfrac{1}{2{,}048}) + \ldots \\ &= \tfrac{1}{2} + \tfrac{1}{2} + \tfrac{1}{2} + \tfrac{1}{2} + \tfrac{1}{2} + \ldots + \tfrac{1}{2} + \ldots \end{aligned}$$

Whoa! Because the series never ends—there is a real, if minute chance that many, many tails are thrown before the first head appears—infinitely many instances of ½ must be summed, and the expected win—lo and behold!—amounts to infinity.

This is really curious, very, very curious. Could Paul, by entering the game, expect to gain an infinite amount of money? Montmort reacted with incredulity. "I am not able to believe that in the case that one gives to Paul some coins according to the sequence 1, 2, 4, 8, 16, 32 etc., the advantage to Paul would be infinite." This raises the question: How much should Paul be willing to pay to participate in such a game? Using common sense, one would think that a gambler would be willing to pay any amount just below the expected win to enter a game. If an expected win is $20, the gambler should be prepared to pay, say, $19.50. But what if the expected win is infinite? Would Paul be willing to pay an infinitely large fee to participate in this game? Or $100,000? Or $10,000? Would you, the reader, offer even $100?

The answer is most certainly no. Nobody would offer more than a few bucks to play the game. But the evidence presented by Nikolaus is irrefutable: The expected winnings are infinite. Hence, everybody should be willing to pay an enormous fee to participate in the game and still expect to strike it rich. Anybody should... but nobody *would*. We have a paradox.

Deeply intrigued, Montmort intended to do some serious thinking about the problem. But he soon realized that his efforts were not evolving into anything worth reporting: "I have not the strength to begin such a great task; this will be for another time." Two years later, the work was still not done, even after prompting by Nikolaus. "I ask you for some time because I am absentminded and lazy," Montmort responded. "Some time" became lots of time, weeks turned into months, and months into years. It would be more than a decade until the matter was taken up again... by someone else.

The person who did so was a twenty-four-year old mathematician named Gabriel Cramer (figure 1.2). As he hailed from Geneva, he could be considered a compatriot of the Bernoullis, except that the Republic of Geneva would become part of Switzerland only in the following century.

Montmort had published the second edition of his *Analyse, revue et augmenté de plusieurs lettres* (Analysis, revised and augmented by several letters) in 1713, to which he appended the correspondence with Nikolaus Bernoulli. Cramer, who diligently read everything that was published on the theory of probability, familiarized himself with the question that Nikolaus had posed and came up with an ingenious explanation for the paradoxical situation. On May 21, 1728, writing from London, he sent a letter to Nikolaus. "I believe that I have the solution of the singular case that you have proposed to Monsieur de Montmort," he wrote.

He was still a little hesitant about his proposal, which is why he prefaced his explanations with a cautionary note: "I know not whether I deceive myself." Obviously, it seems absurd, he wrote, that anybody would be willing to pay an infinite amount of money to enter into such a game "since no person of good sense would wish to give [even just] 20 coins." The reason that Cramer gave for the discrepancy is that there must be a difference between the mathematical calculations and what he calls the "vulgar estimate". "Mathematicians estimate money in proportion to its quantity, and men of good sense in proportion to the usage that they make of it."

FIGURE 1.2: Gabriel Cramer.

Source: Wikimedia Commons

This was a far-reaching insight. It was also quite surprising. How could Cramer (who was, after all, a mathematician) claim that a number was not what it represented? "A rose is a rose is a rose," as the poet Gertrude Stein wrote, expressing that things are just what they are. And now, all of a sudden, a ducat should no longer be a ducat? This is exactly what Cramer claimed. It was an enormous jump. But it would prove crucial, and, as will be recounted in this book, it is the very basis of all economic behavior.

To illustrate his point, Cramer argued that any amount beyond $20 million does not add any utility to men of good sense. Thus, $20,000,001 is no more valuable than $20,000,000, he claimed. In a way, this conforms to everyday experience. Adding $1 to a multimillionaire's bank account will not make her any happier, in the same way that another cookie gives no additional pleasure, even to gluttons, after they have already consumed a dozen. Paupers value a single dollar differently than the rich do.

If one accepts Cramer's reasoning about people's tastes for money, the conundrum posed by Nikolaus Bernoulli is easily resolved. Let us see how: To simplify the calculations, Cramer assumed that any amount beyond 2^{24}, which is close enough to 20 million, gives no additional utility to its owner. It is as if his win remained at 2^{24}. The calculation is then as follows:

$$\begin{aligned} \text{Maximum fee} &= \text{Expected win} \\ &= (1 \times \tfrac{1}{2}) + (2 \times \tfrac{1}{4}) + (4 \times \tfrac{1}{8}) + \ldots + (2^{24})/(2^{25}) \\ &\quad + (2^{24})/(2^{26}) + (2^{24})/(2^{27}) \ldots \\ &= (24 \times \tfrac{1}{2}) + (\tfrac{1}{2} + \tfrac{1}{4} + \tfrac{1}{8} + \ldots) \end{aligned}$$

Because $\tfrac{1}{2} + \tfrac{1}{4} + \tfrac{1}{8} + \ldots = 1$, the expected win is $12 + 1 = \$13$ which, Cramer claimed, is what one should be willing—at most—to pay to enter the game. This is a much more reasonable fee than the infinite amount that the raw calculations suggest. Even though $13 may be somewhat on the high side, it accords more or less with what one would be willing to spend. So, Cramer was on the right track. But he soon realized that his reasoning contained a flaw. It simply cannot be true that any amount beyond $20 million—whether large or small—gives no pleasure at all. Even millionaires would acknowledge that they can get more pleasure from $100 million than from $10. Hence, Cramer realized, an additional dollar could not lack utility entirely, neither to paupers nor to the rich. His theory needed an overhaul.

He quickly found a fix. While the utility of an additional dollar should be less than the utility of the previous one, it should not be nothing at all. In other words, the utility of an additional dollar should be more than zero, but diminish as wealth increases. Thus, as an indicator of the wealth's utility, Cramer suggested the square root of wealth. The ingenious idea behind his suggestion was that when one looks at a graph of the square root function, one notes that it is curved toward the *x*-axis. It grows and grows and grows, but it does so at a diminishing rate. This is exactly what Cramer was looking for. Because the square root of 1 is 1 and the square root of 2 is 1.4142 . . ., the second dollar is valued at about 41 cents. And the ten million and first dollar would be valued at about 0.016 percent of the very first dollar's utility. A rose is a rose is . . . the square root of a rose.

With this new scheme, Cramer recomputed the fee that one should be willing to pay to enter the gamble:[6]

Maximum fee = Expected win = ½ × $\sqrt{1}$ + ¼ × $\sqrt{2}$ + ⅛ × $\sqrt{4}$ + . . . = 2.91 . . .

While $13 may have been more than one wants to spend to enter the game, a little under $3 may be too little. So, depending on the person's personal preferences, Cramer's numbers may be off somewhat, but the principle was definitely correct. The mathematician from Geneva had hit on what would turn out to be the holy grail of economic behavior.

Nikolaus Bernoulli read Cramer's letter with interest, but he was not about to go overboard. Instead of applauding the deep insight, he dismissed it: "The response you give . . . suffices, as you say, to show that A must not give B an infinite equivalent. But it does not demonstrate the true reason for the difference that there is between the mathematical expectation and the common estimate." Oh, indeed! Showing that A must not offer B an infinite sum is what the correspondence was all about. Thus, to belittle Cramer's explanation was very unfair. And as concerns the "true reason," why would Cramer's explanation not be the true reason? Why would the reason that Nikolaus was about to give be any better than Cramer's?

The glib reaction seems all the more unfair because it may have been Cramer's very idea that had struck a chord in Nikolaus's mind in response. Nikolaus maintained that the reason for the paradox was not that amounts beyond a certain large sum give little or no utility to a gambler, but that the gambler disregards minute probabilities. A tiny probability of winning, even if the gain would be very large, is ignored. Hence, he sets all probabilities beyond 1/32 to zero:[7]

$$\begin{aligned}\text{Maximum fee} &= \text{Expected win} \\ &= (\tfrac{1}{2} \times 1) + (\tfrac{1}{4} \times 2) + (\tfrac{1}{8} \times 4) + (\tfrac{1}{16} \times 8) + (\tfrac{1}{32} \times 16) \\ &\quad + (0 \times 32) + (0 \times 64) \ldots \\ &= 2.5\end{aligned}$$

In short, according to Nikolaus, a gambler considers any sequence that starts with five or more tails before a head is thrown (TTTTTH . . .) as so rare that he completely disregards the possibility of such an occurrence. The difference between the two men's explanations was that Cramer assumed that a gambler would set the utility of any amount beyond 2^{24} to zero while Bernoulli postulated that he would set any probability less than 1/32 to zero.

Why should the latter explanation be truer than the former? Instead of substantiating his reservations, Nikolaus took the easy way out. "There may well be some things to say on this matter, but not having the leisure to arrange or to develop the ideas which are presented to my spirit, I pass over them in silence." Do you recall Montmort's lament that some authors always have a thousand things to say that one could very easily do without? This time, one can only regret that Nikolaus did not expend another 100 or 200 words to explain why he condescendingly dismissed his young colleague's suggestion. Apparently he did not realize, or could not get himself to admit, that Cramer had hit upon a very profound idea.

Cramer was then in the town of Leiden, in Holland. In his return letter, he struck a contrite tone: "I have not pretended to guess what the reason is that urges a man not to grant an infinite amount. I have only wished to seek a reason to persuade myself that I must not give an infinite amount." He defended his particular approach to the problem by arguing that it is easier to determine, by introspection, the degree of riches beyond which an additional coin offers no utility than it is to decide at what point a probability should be ignored. As academic disputes go, Cramer's argument is just as legitimate as Bernoulli's, if brought forth in a more respectful manner.

Nikolaus realized that he could not convince Cramer and decided to co-opt Daniel Bernoulli, another of the family's prodigies. In a letter to his cousin, younger by thirteen years, he described the problem and mentioned that Professor Cramer from Geneva had already proposed a solution. But he divulged no details, instead asking for Daniel's opinion. Daniel took up the gauntlet, discussing the problem with his father, Johann. Barely three weeks later, he acknowledged the existence of a paradox. Like his cousin, he believed that the answer to this paradox lay in the exceedingly small probabilities that a game would last more than twenty or thirty throws.

As time went by, Daniel was not quite satisfied with his answer. In a further letter, now lost, he raised another interesting issue. We may surmise from Nikolaus's displeased answer that Daniel had come to accept Cramer's point of view: Player B, who must eventually pay the winning amount to player A, may not be rich enough to pay the 17 million coins that he would owe A if the first heads appeared after twenty-five consecutive throws. Hence, it is no wonder that player A values any winnings beyond 2^{24} as zero. He would never get more than that amount anyway because his opponent does not possess that much money.

Nikolaus dismissed his cousin's arguments, as he had done with Cramer's, and for nearly a year, Daniel did not react. Finally, in January 1731, a letter arrived on Nikolaus's table. Daniel insisted that the other player's inability to pay an extremely large reward was crucial to the understanding of the paradox. "I have no more to say to you, if you do not believe that it is necessary to know the sum that the other is in a position to pay." After all, he continued, "a person would not wish to play against another who [is willing] to wage an infinite sum in a game, [even if] there is [only] an infinitely small probability."

Now it was Nikolaus who kept silent for several months. Frustrated, Daniel put down his ideas in a manuscript to which he gave the title *Specimen theoriae novae metiendi sortem pecuniariam* (Exposition of a new theory of measuring financial outcomes). Because he taught in St. Petersburg, Russia, together with his learned friend from Basel, Leonhard Euler, he presented the paper at a session of that city's Academy of Science. Then, on July 4, 1731, he sent the manuscript to Nikolaus.

Once again, the elder cousin was perturbed. "Most ingenious," he snorted, and quickly added his reservation: "Permit me to say that it does not solve the knot of the problem." With a convoluted argument why, in his view, it did not, he dismissed an important observation by both Cramer and Daniel. They had tried to convince Nikolaus that additional ducats afford less utility than the previous ones using the argument that "the pleasure or advantage of a gain in the favorable case is not equal to the sorrow or disadvantage that one suffers in the contrary case." In modern parlance, a gain of $10 gives less pleasure than a loss of $10 hurts. What Daniel and Cramer proposed (and Nikolaus discarded) was, once again, nothing less than the basic principle on which all economic behavior is built. Alas, Nikolaus remained adamant.

Nevertheless, the argument that a player might be unable to settle his debt struck a chord with Nikolaus again. It also gave him an opportunity to change the subject. Turning away from coin throws and toward financial investments, he suggested that it is preferable to split a sum of money among several people rather than be exposed to a single debtor's ability to make good on his promise. "One nevertheless does better to place 500 coins in two places, than 1,000 coins in a single place, because one is not exposed to losing as easily all 1,000 coins in the first case as in the second." With each debtor having a 10 percent probability of defaulting, Nikolaus explains, there is a 10 percent chance of losing all one's money if it is placed with a single debtor, while there is only a 1 percent chance of losing all the money if it is placed with two debtors.[8] To buttress his contention, he quotes the familiar proverb, apparently already known to the burghers of Basel in the eighteenth century: "One must not put all eggs in one basket." So, while he discarded what would become a basic tenet of economic theory, he formulated what would become one of the basic tenets of financial theory.

The manuscript that Daniel had sent to Nikolaus was only a draft. Over the next seven years, he sharpened the arguments, polished the text, and created a new version of the document. When he was satisfied that it was sufficiently elegant to be presented to the scientific public, he submitted the eighteen-page paper to the Proceedings of the Imperial Academy of Sciences in Petersburg. It was published in 1738, in Latin, under the title *Specimen theoriae novae de mensura sortis* (Exposition of a new theory on the measurement of risk). To this day, it is considered one of the outstanding academic articles in economics—so much so that more than two centuries later, in 1954, the prestigious journal *Econometrica* published an English translation of the often-cited but little-read paper.

Right from the start, in section 1 of his booklet, Bernoulli took issue with the common wisdom that the expected value of a gamble is computed by multiplying each possible gain by its probability and summing the products.

If this were exactly so, he wrote, everything would be easy. One should be willing to enter into the gamble by paying as much as its expected value. The personal circumstances of the decision-maker—rich or poor, optimistic or pessimistic, happy or sad—would play no role; mathematical rules would govern the decision process, and all people would agree on the correct choices to be made. There would be "no need for judgment but [just] of deliberation."[9]

But this is obviously not the case. Bernoulli illustrates this point with a lottery ticket that yields either nothing or 20,000 ducats with equal probability, and he proceeds to tell the story of a poor man and a rich man who would make different decisions when presented with this situation. The pauper would be well advised to sell such a lottery ticket for 9,000 ducats, Bernoulli recommends, even though the mathematical expectation is 10,000. The rich man, on the other hand, would be well advised to buy it for 9,000 ducats. "It seems clear that not all men can use the same rule to evaluate the gamble. The directive established in paragraph §1 must, therefore, be discarded." This insight led straight to his main point—namely, that "the value of an item must not be based on its price, but rather on the utility it yields. The price of the item is dependent only on the thing itself and is identical for everyone; the utility, [however], is dependent on the particular circumstances of the persons making the estimate."[10] For those readers who still do not get it, he hammers in the obvious: "A gain of one thousand ducats is more significant to the pauper than to the rich man, though both gain the same amount."

As a consequence, the instruction in section 1 must be corrected. Instead of multiplying values by probabilities, Bernoulli states, it is the utility of each possible profit that must be multiplied by its probability. Thus, one obtains a *mean utility*, which then needs to be converted back to its corresponding monetary value. This is the lottery's value to the person.

The question now was how a person determines utility of wealth. Bernoulli makes a reasonable proposal: "The utility resulting from any small increase in wealth will be inversely proportional to the quantity of goods already possessed."[11] The more that one has, the less is one's utility for additional wealth. Of course, there are exceptions, Bernoulli concedes, but the principle is sound.

For the mathematically inclined, here is an aside. According to Bernoulli, if one owns wealth W, which gives utility $U(W)$ to its owner, and this owner stands to gain a small additional amount of money, dW, then the additional utility $dU(W)$ is

$$dU(W) = c\, dW/W,$$

where c is a parameter that may differ from one individual to another. Hence,

$$dU(W)/dW = c/W,$$

which, after integration, results in

$$U(W) = c\, Logarithm(W) + constant.$$

Thus, if one accepts Bernoulli's assertion, utility of wealth follows the logarithmic function. It is a reasonable suggestion because that function always rises, as one expects the utility of wealth to do, but an additional ducat always provides less utility than the previous one. Bernoulli's reasoning requires a leap of faith, however, as he provides no proof that additional utility is in fact inversely proportional to current wealth. Nevertheless, the logarithmic utility function is an improvement over Cramer's square root utility function, which had nothing going for it other than its shape. At least, the logarithm is not only convenient, but it also has a sensible (albeit unproven) underpinning... and it is elegant.

Let's illustrate this point with an example. It is a little different from the one Bernoulli presented in his paper, in that it is a little less complicated. Monsieur Pimpodou, who possesses 1,000 ducats that he wants to invest, is presented with a business proposal that promises either to double or to triple his stake, with equal probability. If he accepts the venture—and who wouldn't?—he would end up with a final tally of either 2,000 or 3,000 ducats. Hence, mathematically, the expected profit would be 1,500 ducats (50 percent of 1,000 + 50 percent of 2,000). This means that Monsieur Pimpodou, if he is rational, should be willing to give up his right to invest in the proposal in exchange for an immediate payoff of 1,500 ducats.

Let us see how Bernoulli goes about analyzing this situation. According to his proposal, the relevant variable that must be considered is not the value of the person's final wealth positions in ducats but rather the utilities of these wealth positions (i.e., the logarithms of their ducat values). Because the logarithm of 2,000 ducats is 7.6 and the logarithm of 3,000 ducats is 8.0, the *expected utility* of Monsieur Pimpodou's final wealth position is 7.8 (50 percent of 7.6 + 50 percent of 8.0). And what amount of wealth would give him just this utility? The answer is 2,440 ducats because the logarithm of 2,440 is 7.8.[12]

Thus, Monsieur Pimpodou would be just as happy to add 1,440 ducats to his existing wealth of 1,000 ducats as he would be to participate in the business venture. But 1,440 is less than 1,500, the sum for which a rational person would be willing to settle. What's going on? Bernoulli asserts that when the shrewd Monsieur Pimpodou considers his utility, he prefers to receive 1,440 ducats for sure than to face the uncertain prospect of receiving either 1,000 or 2,000 ducats. He is willing to give up 60 ducats from the mathematically expected value of the venture (1,500 − 1,440) in order to get a little less, but to get it with certainty. And therefore, with hardly any ado, Bernoulli introduced a principle of such paramount importance that it is still with us 300 years later and will remain a defining concept of economic behavior forever: Pimpodou is averse to risk! He does not act according to the dry analysis of mathematical expectation, but Pimpodou is not irrational at all. He acts according to the perfectly logical assumption that an additional cookie gives less utility, even to the glutton, after he has already consumed a dozen of them.

The importance of the principle just described here cannot be overstated: *The fact that an additional ducat gives less utility to a rich man than to a pauper entails that people are generally averse to risk.*

For the less mathematically and more geometrically inclined, I illustrate Bernoulli's suggestion, as he did as well, with a graph (see figure 1.3). The horizontal axis, which we shall call the *$-axis*, denotes ducat values, and the vertical axis, the *U-axis*, designates the utility. The bent curve, which is the graph of the logarithmic function, shows how Monsieur Pimpodou translates dollars into his utility and vice versa. This curve always rises, but it is bent downward throughout, which indicates that each additional coin affords him less utility than the previous ones. Obviously, this means that the same additional sum gives less utility to a rich man (A) than to a pauper (B).

Let us take this scenario step by step. First, we indicate a wealth of 2,000 ducats on the *$-axis* (*a*). We trace this value to the utility curve (*b*), which

FIGURE 1.3: **The utility of wealth.**

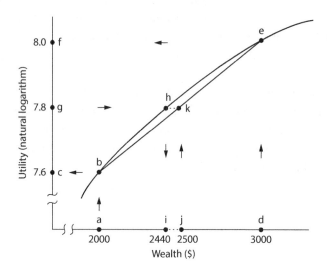

Note: the curve (through points *b*, *h*, and *e*) is the natural logarithm; that through points *b*, *k*, and *e* is simply a straight line.

indicates a utility on the *U*-axis of 7.6 (*c*). We do the same for 3,000 ducats, on the $-axis (*d*), tracing it to the utility curve (*e*), which indicates a utility of 8.0 on the *U*-axis (*f*). Now we seek the expected utility on the *U*-axis, which is halfway between 7.6 and 8.0 [i.e., at 7.8 (*g*)]. (It is halfway because the odds are 50-50. If the odds were different, the location on the *U*-axis would have to be adapted accordingly.) The question is: What ducat value corresponds to a utility of 7.8? Let us trace from the expected utility level of 7.8 toward the utility function (*h*), which indicates a value of 2,440 on the $-axis (*i*). However, the mathematically expected monetary value of the gamble lies halfway between 2,000 and 3,000 ducats [i.e., at 2,500 ducats (*j*)]. Now we are done. A final, certain wealth value of 2,440 ducats gives Monsieur Pimpodou the same utility that he would expect to enjoy from gambling on the proposal. The monetary difference between *i* and *j*, 60 ducats, is the price that he is prepared to pay to avoid the gamble.

We now see, in black and white, why the fact that an additional ducat gives less utility to a rich man than to a pauper leads directly to risk aversion.

The curve that describes the utility for wealth slopes upward, but it is also curved downward—in technical terms, the curve is concave—and that means that the owner of this utility function always wants to forgo a premium in order to avoid risk.

The consequences of what we glean from this simple graph are enormous. Daniel Bernoulli mentions two of them. The first is obvious: Forgoing money to avoid a risk is equivalent to buying insurance against uncertainty. Actually, insurance flourished long before Bernoulli gave the behavioral underpinnings for the industry's existence. Babylonian traders as far back as the third millennium BCE did not need to be told about utility and logarithms. They knew intuitively that it was correct to give up some money to avoid uncertainty.

However, a question arises immediately: Who would want to sell insurance? An insurance company—run by risk-averse people like Monsieur Pimpodou—also dislikes uncertainty. So why would it be willing to assume the same risk that Monsieur Pimpodou is trying to get rid of? The reason, Bernoulli says, lies in the differing wealths of the decision-makers. An insurance company is much richer than an individual person, and therefore it demands a smaller risk premium. Let us say that the company owns 100,000 ducats in assets. Repeating the previous calculations, translating ducats into logarithmic utility and back again, the company's risk premium comes to only about 3 ducats. Hence, with Monsieur Pimpodou being prepared to pay 60 ducats, an insurance company—or a very wealthy individual, for that matter—could come to an agreement with him. Of course, an even larger insurance company would willingly take the smaller company's 3 ducats in exchange for the assumption of the risk because the risk premium that it demands would be even smaller. And so it continues . . . The reinsurance business is born.

The other consequence that Bernoulli mentioned in his treatise has less to do with risk premiums as such and more with diversification. If a merchant transports all his goods on one boat, and one in ten boats sinks, there is a 10 percent chance that the entire fortune would be lost. Bernoulli advises distributing the goods evenly between two boats. Then, the probability that both boats would sink and that the merchant would lose everything, is reduced to 1 percent. (The probability that *both* boats will sink is 10 percent × 10 percent, or 1 percent.) There is, unfortunately, the additional possibility that one of his boats will sink and the merchant will lose half his wares. The probability of this happening is 18 percent (90 percent that Boat 1 gets through multiplied by 10 percent that Boat 2 sinks, plus 90 percent that Boat 2 gets through multiplied by 10 percent that Boat 1 sinks). Using the by-now familiar utility calculations, Bernoulli showed that, utilitywise, a 1 percent chance of losing all goods, plus an 18 percent chance of losing half, is preferable to a 10 percent chance of losing all. It's "Don't put all your eggs in one basket" all over again, as his cousin had admonished six years earlier.

There is another reason, not mentioned in Bernoulli's paper, that insurance companies are willing to take on risks that Monsieur Pimpodou and his ilk want to avoid. It is because the law of large numbers kicks in. A single ship, with all the trader's wares on board, either arrives at the port of destination or it does not. But with many dozens (or even hundreds) of ships in the company's custody, there is much less uncertainty about the ratio of successful voyages to shipwrecks. Thus, the premiums that all traders are willing to pay will compensate the insurance company for the remaining, much smaller risk.

Bernoulli ends his tractate with a tribute to his cousin's pen pal, Gabriel Cramer. His most honorable cousin Nikolaus, he wrote, "informed me that the celebrated mathematician Cramer had developed a theory on the same subject several years before I produced my paper. Indeed, I have found his theory so similar to mine that it seems miraculous that we independently reached such close agreement on this sort of subject."[13] He proceeded to copy Cramer's missive to Nikolaus, word for word, which is how it was preserved for posterity.

A final remark is in order here. Cramer used the square root function to describe a person's utility for wealth, mostly out of convenience. Bernoulli, with a bit more theoretical justification, opted for the logarithmic function to translate a person's wealth into utility. But the function need not be either the square root or the logarithmic. Any curve that slopes upward but bends downward might do. We will return to this point in later chapters.

It took the foremost mathematicians of their time twenty-five years to wrestle with the problem that henceforth became known as the St. Petersburg Paradox, from Nikolaus Bernoulli's letter in 1713 to Daniel Bernoulli's treatise in 1738. But now the table was set.

We now know that utility for wealth increases, and it does so at a decreasing rate. The next chapters will discuss these two phenomena in more detail.

CHAPTER 2

MORE IS BETTER...

We hold these truths to be self-evident, that all men are created equal, that they are endowed by their creator with certain unalienable Rights, that among these are Life, Liberty, and the pursuit of Happiness.

So it says in the American Declaration of Independence, signed by the Second Continental Congress on July 4, 1776. In this chapter, we leave aside the important concepts of life and liberty and deal mostly with the pursuit of happiness, which has much bearing on this book's subject. There is a debate among political scientists on whether Thomas Jefferson, later the third president of the United States, was influenced by the English philosopher John Locke when he penned these words; or whether one has to go even further back in history, to ancient Greece, to understand Jefferson's thinking on the matter of happiness. For his part, Jefferson considered himself a follower of Epicurus, the fourth-century B.C. philosopher, as he stated in a letter to his private secretary, William Short. So that is where we shall start.

To establish the context, we will examine a predecessor of Epicurus, the philosopher Aristippus, a disciple of Socrates from Cyrene, an ancient Greek city on the north African coast, in what is now Libya. He lived from about 435 to 356 BC. What is known about his life comes mostly from the writings of Diogenes Laertius, a biographer of Greek philosophers who lived about two centuries after Aristippus and Epicurus.

Aristippus heard about Socrates as a young man, and he was so impressed by the philosopher's wisdom that he traveled to Athens to study under him.

Genteel and calm, Aristippus is said to have exuded poise and charisma. When he felt that he had learned enough, he branched out on his own but—horror of horrors—took money from his students for his services. In defense of his tuition policy, he pointed out that it was easy for Socrates to teach free of charge because his followers brought him whatever he required and personally attended to all his needs. He, on the other hand, had to pay good money for his groceries and had to purchase a slave to do housework for him. Once, a father remarked that for the sum demanded, he could buy a slave to instruct his son. Unruffled as always, Aristippus (figure 2.1) responded, "Then do so, and you will have two."

What was it that he taught? The basic doctrine of his philosophy was a very convenient one. Pleasure is what everybody should strive for, Aristippus conveyed to his disciples; it is life's goal. The supreme objective of every human being is to seek pleasure and to avoid pain, to maximize net happiness (i.e., the sum total of all pleasures minus all pains). It is a very laid-back, easygoing principle, indeed. Compare it to the teachings of other philosophers, who maintained that lofty goals like virtue, justice, moderation, and learning should be the purposes of a god-pleasing life. Aristippus would have none of that. For him, happiness was all that mattered.

Like most philosophers worth their salt, Aristippus practiced what he preached. Known as a *bon vivant*, he lived in luxury and indulged in various pleasures. In particular, he sought happiness in the arms of Laïs, a beautiful and capricious courtesan.[1] When one of his disciples reproached him for conducting an affair with this notorious woman, who had bestowed her favors on so many previous lovers, Aristippus responded that he would not decline to travel in a boat in which many had traveled before him, nor would he refuse to live in a house where others had lived earlier. And anyway, he remarked, "it is not the abstinence from pleasure that is best, but mastery over them without being worsted."[2] Alas, Laïs did not return his love. Instead she fell for an athlete, Eubotas of Crete, who had won the foot race and the chariot race at the Olympic Games. (Ironically, her own love went unrequited as well: When Eubotas returned to his home in Cyrene, he refused to let her accompany him, as he had promised, and took along only her portrait.)

Among the pupils Aristippus instructed was his daughter, Arete. (Presumably, in this case he demanded no money from her parent.) She, in turn, imparted

FIGURE 2.1: **Aristippus.**

Source: Wikimedia Commons; in *The History of Philosophy*, Thomas Stanley (1655)

his teachings to her own son, Aristippus the Younger, who perpetuated his grandfather's philosophy that was henceforth known as hedonism. But it was Epicurus (figure 2.2), born in 341 BC, about twenty years after Aristippus died, who is considered his successor—albeit to some extent a contrarian one. The basic tenet of Epicurus's school was that pleasure was the "alpha and omega of a blessed life."[3] This was taught in idyllic surroundings:

> In a garden situated in the outskirts of Athens, a small body of men and women bound together in friendship by similarity of tastes and their belief in a common doctrine, walked and talked, living a simple and natural life, discoursing on philosophy, and letting the great world go on its way. . . . It was an uneventful and leisured life of a small university in which rivalries and ambitions were dissolved in reverence for a loved teacher, and into which no disturbing spirit, burning with a sense of the wrongs and woes of humanity, was permitted to enter. A quiet, dreamy, cloistered life it was, ennobled by an air of antique grace and refinement.[4]

The Epicureans' outlook, however, was more refined than that of Aristippus, which could be seen as advocating a life of total, and immediate, indulgence. "When we say, then, that pleasure is the end and aim, we do not mean the pleasures of the prodigal or the pleasures of sensuality," Epicurus explained in a letter to a friend. "It is not an unbroken succession of drinking-bouts and of revelry, not sexual love, not the enjoyment of the fish and other delicacies of a luxurious table, which produce a pleasant life."[5] No, life should be governed by sober reasoning. While pleasure, both physical and mental, is the "starting point of every choice and of every aversion," he cautioned that "we cannot lead a life of pleasure which is not also a life of prudence, honour, and justice." Not all pleasure is to be sought, especially if it entails harm later on—like a hangover following a drinking binge, or punishment after a transgression—but happiness also can be found in inconspicuous places. Indeed, apart from owning a splendid garden, Epicurus lived a very simple life. Plain bread and water were his usual fare, only occasionally supplemented by a pot of cheese. His followers were no less frugal, even at festivities. According to Diogenes, "At all events, they were content with half a pint of thin wine and were, for the rest, thorough-going water-drinkers."[6] Surprisingly, such were the customs at the school that held that pleasures were the aim of life.

FIGURE 2.2: Epicurus.

Source: Wikimedia Commons; Palazzo Massimo alle Terme

Epicurus's end was horrible. He had suffered from kidney stones throughout his life, until, at age seventy-one, his urinary tract was completely blocked. For two long weeks, he suffered terribly. Feeling that his end was near, he penned one last letter to a close friend, entered a warm bath, and, partaking of a cup of wine, passed away. Speaking loosely, one could say that he accepted his fate stoically, except that stoicism was a different philosophical school altogether.

I have given a very cursory and rather idiosyncratic interpretation of hedonism and Epicureanism. The point I wanted to make is that according to the tenets of Aristippus and Epicurus, the goal in life is to maximize your pleasures; the more of them, the better. Epicurus envisaged two sorts of happiness: "the one, the highest possible, such as the gods enjoy, which cannot be augmented, the other admitting addition and subtraction of pleasures."[7] The latter, the sort of happiness that all of us enjoy, can always be increased. When he admonished his contemporaries to be frugal, remarking that "nothing is enough to someone for whom what is enough is little," he recognized that most humans tend to desire ever more wealth. Thence arose the notion, affirmed in the Declaration of Independence, that all citizens, no matter how rich, have the right to pursue happiness by maximizing pleasure and minimizing harm.

There is a catch, however. Admittedly, consuming a cookie often adds to one's pleasure. But after one has already eaten a dozen, it probably does not. In fact, it may *decrease* pleasure. Hence, a good strategy would be not to consume this new cookie immediately, but to save it for later, when it would again add to one's pleasure. Unfortunately, such a strategy does not hold for ice cream, fish, meat, milk, and other perishable goods.

This is where money comes in. The owner of perishable items can exchange them for money, which in turn can be exchanged for pleasure-adding items again at the appropriate time. Hence, the more money one has, the better. Admittedly, as the saying goes, money cannot buy happiness—but as another saying goes, it helps. Having money allows the purchase not of all, but of additional pleasures. And having more money allows the purchase of more pleasures. One of the first thinkers to consider the role of money was the seventeenth-century physician John Locke (figure 2.3).

Born to a well-to-do family in 1632, a century before the Bernoulli cousins discussed what would become known as the St. Petersburg Paradox, Locke attended the Westminster School in London, which even today is considered one of Britain's leading secondary schools. Middle school was followed by the

FIGURE 2.3: John Locke.

Source: Wikimedia Commons

study of classical languages, logic, and metaphysics at Oxford. But the curriculum bored Locke, who would have preferred to study modern thinkers like René Descartes instead of the works of Plato and Aristotle. But he persevered, obtaining a bachelor's degree in 1656, a master's degree in 1658, and, following a long-lasting interest, a belated bachelor's in medicine sixteen years later. In 1668, Locke was admitted as a Fellow to the Royal Society, where he was joined by Isaac Newton four years later. For most of his life, he was relatively independent financially because he had inherited some property upon his father's death.

Actually, the medical degree had been long overdue—because, while still a student, he had already successfully treated Sir Anthony Ashley-Cooper, later the Earl of Shaftesbury and Lord Chancellor, for a life-threatening liver infection. So impressed was Sir Anthony by Locke's skills that he offered him employment as his personal physician and secretary. As his patron's fortunes rose and declined, Locke went in and out of favor too. When he came under suspicion of plotting against King Charles II in 1683, he had to seek refuge in the Netherlands, returning to England only six years later, together with Mary, Princess of Orange, later to be Queen Mary. Through diplomatic service in Germany, travels through France, exile in the Netherlands, and as commissioner of the Board of Trade and Plantations (1673–1674) and a member of the Board of Trade (1696 to 1700), Locke gained firsthand knowledge and extensive experience in politics, governance, economics, commerce, administration, and international trade.

Throughout his life, Locke wrote on political theory, religion, economics, education, and consciousness. He wrote his most important theses on political philosophy, *Two Treatises of Government*, during the dark days of Charles II's drift toward absolutism, and published the tome anonymously in 1689, just after his return from the Netherlands. This groundbreaking work was reprinted several times during his lifetime (though never to his satisfaction) and is considered by today's scholars to be the foundational text of liberalism. It proposed a theory of natural law and natural rights that Locke used to differentiate between legitimate and illegitimate governments. With absolute monarchy being one of the latter, he endorsed the right of resistance to tyrants.

Always cautious, Locke took great pains to keep his authorship secret—and he had good reason to do so. The politician Algernon Sidney, a like-minded thinker on political theory, was convicted as a traitor and executed based on his as-yet-unpublished manuscript, *Discourses Concerning Government*.[8] Confiding only in his most trusted acquaintances, Locke destroyed all his manuscripts and expunged any reference to his authorship in his papers. All negotiations with printers and publishers were conducted via third parties. Nevertheless, his authorship soon became the subject of rumor, if not certain knowledge, among the English intelligentsia.

I will not delve into what Locke had to say about government; instead, I limit my remarks to his observations about private property, accumulation, and money. Locke's basic tenet is that each individual owns at least himself. Obviously, this means that his body belongs to him and that all the labor he performs also belongs to him. On the other hand, the produce of the earth was given by God to all humanity, or put another way, belongs to nobody in particular. Only when an individual adds his or her labor to a natural resource, it comes into his or her possession. If, for example, a woman picks an apple from a tree, the apple then belongs to her. Without her labor, the apple has no value. By mixing one's labor with the natural materials that the earth gives to humanity, one creates food, clothes, and shelter (hence one's property). The acquisition of property and its accumulation without limit, the just fruits of a person's labor, are endorsed as a legitimate endeavor by Locke. In fact, "the chief end [of civil society] is the preservation of property," he remarked.[9]

There is a caveat, however. Many natural resources are scarce. And many can spoil. One is not allowed to let one's accumulated property go to waste, Locke admonished. An individual who hunts animals or gathers vegetables must make sure "that he used them before they spoiled, else he took more than his share, and robbed others." To pick more apples or gather more acorns than one can eat before they spoil would be a transgression against one's fellow-man, who, after all, is a coproprietor of the produce before it goes to waste. For an individual, it is "a foolish thing, as well as dishonest, to hoard up more than he could make use of"[10]

The conclusion thus far is that one may accumulate all that one is able to use for personal consumption, but not more than that. However, it may require only part of the family's time to hunt for sufficient meat, gather enough produce, and harvest adequate amounts of crops for oneself and one's family. So how can one use one's ability to expend more labor, and thus acquire additional property? Locke gave the answer: "And thus came in the use of money, some lasting thing that men might keep without spoiling, and that by mutual consent men would take in exchange for the truly useful, but perishable supports of life." It is the invention of money that made accumulation possible . . . and legitimate. "And if he also bartered away plums, that would have rotted in a week, for nuts that would last good for his eating a whole year, he did no injury; he wasted not the common stock . . . If he would give his nuts for a piece of metal, pleased with its colour; or exchange his sheep for shells, or wool for a sparkling pebble or a diamond, and keep those by him all his life, he invaded not the right of others."[11]

Locke recognized that once money had been invented, and with human nature being what it is, people would eagerly take the opportunity to increase their wealth: "Find out something that hath the use and value of money amongst his neighbours, you shall see the same man will begin presently to enlarge his possessions." Acquisitiveness, possessiveness, and even greed were just part of the normal state of affairs because every person "might heap up as much of these durable things as he pleased." In Locke's opinion, differences in income and wealth were entirely justified because "as different degrees of industry were apt to give men possessions in different proportions, so this invention of money gave them the opportunity to continue and enlarge them . . . ." "It is plain," he summed up, "that men have agreed to a disproportionate and unequal possession of the earth, they having, by a tacit and voluntary consent, found out, a way how a man may fairly possess more land than he himself can use the product of, by receiving in exchange for the overplus gold and silver, which may be hoarded up without injury to any one."[12] So Locke too, like Aristippus and Epicurus, was of the opinion that more of a good—or, if the good is perishable, more money—is better than less of it.

With the problem of goods' spoilage taken care of through the introduction of money, Locke needed to find and justify ways to safeguard that money and any further property from thieves, swindlers, and other crooks. Nowadays, we take it for granted that the judicial system and enforcement agencies ensure everybody's property rights. But Locke was exploring the ideas of how societies should live *ab initio* from the very beginning. In *Two Treatises on Government*, he stated that before a government exists, men are in a "state of perfect freedom to order their actions, and dispose of their possessions and persons, as they think fit"[13].

This presents a conundrum, however. If people are totally free, and they have a natural right to such freedom, why would they willingly submit to a government, thus giving up part of their freedom? The answer that Locke proposed sheds light on the importance of money and wealth. A man may possess property, Locke explained, but his enjoyment of it would always be at risk, threatened by unsavory characters and practices. Hence, he concluded that people are willing to join in society with others "for the mutual preservation of their lives, liberties and estates." The role of government is not just to guarantee life and liberty but, significantly, to safeguard property: "The great and chief end, therefore, of men's uniting into commonwealths, and putting themselves under government, is the preservation of their property."[14] Locke died in his study room in 1704. (Daniel Bernoulli was then four years old.)

In 1671, John Locke, in his capacity as medical doctor in Lord Shaftesbury's household, assisted at the birth of the lord's grandson, who was named—like his father and grandfather—Anthony Ashley-Cooper. Later, upon the death of his father, he would become the third Earl of Shaftesbury. Locke, the person most trusted by the first earl, was put in charge of little Anthony's education. He devised a curriculum for the boy and, quite unusual for the time, entrusted day-to-day instruction to a female tutor, Elizabeth Birch, a woman perfectly fluent in ancient Greek and Latin. At age eleven, her charge was also fully conversant in these dead languages.

After a two-year grand tour of Europe, and after the death of his father in 1699, the third Earl of Shaftesbury was put in charge of the family business, which included supervising the education of his brothers, arranging the marriages of his sisters, overseeing the family investments, governing the family lands, and adjudicating disputes among the tenants. In addition, following in the footsteps of his forebears, he became a politician, and at the age of twenty-four, entered Parliament. A lifelong sufferer of asthma or tuberculosis, he was forced to give up his seat three years later and leave smoke-polluted London. For a while, he retired to the Netherlands. Like the first Earl of Shaftesbury, but unlike the second (who was considered, maybe unfairly, a bit of a dunce), the third earl also became a philosopher. In contrast to some of his peers, readability of his work was an important concern for him, and he experimented with various literary forms, and even illustrations, to make his points. His three-volume work *Characteristicks of Men, Manners, Opinions, Times*, first published in 1711, was the second-most reprinted book in English during that century, just behind Locke's *Treatises*. He would not live to see much of the success of his tome. Barely two years after publication of the first edition, Anthony, the third Earl of Shaftesbury, died at age forty-two.

In his writings, the earl emphasized that the virtuous person should, through his actions, contribute to the good of the community. Hence, the advisability of a person's actions lies in whether they promote the general welfare of humanity, and one needs to observe one's impact on the community in which one lives. Although Shaftesbury is not counted among the best-known thinkers nowadays, his writings influenced subsequent philosophers deeply, notably Francis Hutcheson (1694–1748), who in turn inspired Jeremy Bentham (figure 2.4), thus spawning the philosophical school of what would become known as *utilitarianism*.

Bentham, the son of a prosperous clerk at one of London's trade associations, was born in 1748. He was recognized as a prodigy as a toddler, when he was found one day at his father's desk, reading a multivolume history of England. At age three, he began to study Latin, and after attending the Westminster School, as Locke had done thirty years earlier, he was sent to Oxford at age twelve.

Bentham was unhappy at both institutions. A small and frail boy, he was subjected to bullying at Westminster, and he abhorred the school's strict regime and the constant threat of physical punishment by caning. At Oxford, he was dismissively called the "philosopher" by the sons of squires and noblemen who attended Queen's College with him. The burlier among them delighted in showing off their physical (if not intellectual) prowess by holding him upside down. Once he lost his whole allowance during such an ordeal, when a half-guinea fell out of his pocket. But Bentham was unhappy not only because of the tribulations of college life, but also because of the dismal education he received at Oxford. He had nothing but contempt for "the foolish lectures" of his tutors that he was forced to attend. (Ten years later, his younger brother Samuel, to whom he stayed close until the latter's death one year before his own, would profit from educational reforms that his older brother advocated. Samuel became a noted engineer and naval architect.)

Bentham's father, a loving but proud and ambitious man, wished for Jeremy to become a lawyer. He was convinced that he would one day become Lord Chancellor, but the younger Bentham soon became disillusioned about the state of legal affairs in England. The premature end to his career as a barrister came when he counseled a prospective client that litigation would cost more than he stood to gain. So, instead of practicing law, he set about to reform it.

From then on, and until the very last days of his life at age eighty-four, Bentham wrote copiously on all aspects of jurisprudence and political systems,

FIGURE 2.4: Jeremy Bentham.

Source: Wikimedia Commons; painting by Henry William Pickersgill

producing an average of fifteen handwritten folio pages a day. Although loathe to publish the product of his labors, he never doubted its, or his own, importance. Even when confiding his shortcomings, he paid himself underhanded compliments, as when he asserted that timidity, awkwardness, embarrassment, and false shame are "the most cruel enemies of merit and solitary genius."[15] But his self-doubts and twinges of conscience were real. Only after prompting by friends did he consent to publish his work, and even then, solely on the condition that it stay anonymous . . . at least until his delighted father could no longer contain himself and gave away the secret.

Bentham fell in love several times during his long life, but never married. Above all, he seems to have been in love with himself, as witnessed by the bizarre instructions in his will: After his death, his body was to be dissected, and the skeleton dressed in his clothes and preserved. To this day, the "auto-icon" is kept on display in a wooden cabinet at University College London (UCL), a place that Bentham spiritually founded. Only the corpse's mummified head was replaced by a wax replica after several disrespectful pranks by UCL students.

Taking his cue from forerunners as far back as Aristippus, Bentham postulated that the behavior of humans is governed by two guiding concepts: pain, which they try to avoid; and pleasure, which they strive to increase. As a proponent of societal reform, he sought a way to reconcile such egoism with altruism, so that blatant self-interest could be channeled toward behavior that would serve the greater good of society. Defining happiness as the sum total of pleasure minus pain, he came up with what was henceforth called the *greatest happiness principle*. "It is the greatest happiness of the greatest number that is the measure of right and wrong," he wrote in the preface to *A Fragment on Government*.[16] Hence, it is not just one's own happiness, but the happiness of all members of the community that must be taken into account. When considering the advisability of an action, an individual's moral compass should point not toward decency of intentions, like justice, fairness, or equality, but to the utility of the action's outcome to the greatest number of people. Thus, utilitarianism was born.[17]

The maxim "greatest happiness of the greatest number" is opposed to notions of justice or a person's natural rights. In Bentham's system, lying, cheating, and stealing are legitimate acts if such behavior increases the community's aggregate happiness. "Natural rights is simple nonsense: natural and imprescriptible rights, rhetorical nonsense—nonsense upon stilts."[18] The only reason why justice should prevail and be rewarded at all is because it is beneficial for the community in the long run. Bentham opposed the allegedly self-evident, inalienable rights mentioned in the Declaration of Independence. And he minced no words: "The opinions of the modern Americans on government, like those of their good ancestors on witchcraft, would be too ridiculous to deserve any notice."[19]

Why does the Declaration (in particular its preamble) call for such harsh words? Bentham contended that the institution of government is inconsistent with the existence of unalienable, natural rights: "to secure these rights, they [modern Americans] are content that Government should be instituted. They perceive not . . . that, consequently, in as many instances as Government is ever instituted, some or other of these rights, pretended to be unalienable, is actually alienated."[20]

Well, duh! Of course, government forces citizens to obey its laws. After all, laws are instituted for the good of all citizens. If justice should prevail for the greater good of the community, even if it curtails the freedom of the individual, then surely the same can be said for obeying laws and paying taxes. These constraints do not diminish citizens' rights to life and liberty, nor do they diminish their right to pursue happiness.[21]

As the quarrel about the Declaration of Independence shows, the problems associated with the principle "greatest happiness of the greatest number" are too numerous to list. For example, one cannot always know in advance whether an action's unintended consequences will increase or decrease happiness. Is one allowed to utilize amoral means to attain greater happiness for more people? For how many people? Above all, there is one problem about which we will have a lot more to say in later chapters: How can one compare one individual's happiness with that of another? Is the owner of a house allowed to plant a tree in her garden if its shade increases her pleasure, but falling leaves increase the neighbor's pain? On an even deeper level, what are the factors that contribute to, or deduct from, happiness? Bentham thought deeply about that question and then proposed an algorithm, the "felicific calculus," to quantify happiness. It is meant to measure utility mathematically in order to determine whether an action is advisable or not.

Bentham stipulated that happiness is a compound emotion made up of basic pleasures, of which there are fourteen—the pleasures of sense, wealth, skill, power and so on—and pains, of which there are twelve—privation, enmity, awkwardness, an ill name, and so on. He proposed to measure this emotion along seven dimensions. The first four—intensity, duration, probability of occurrence, and propinquity (how soon will it occur?), are associated with the sensation of pleasure or pain itself that the individual experiences. The next two—fecundity (will more of the same sensation follow?) and purity (will the sensation be followed by the opposite sensation?)—measure the consequences of the action. The final dimension, the extent, measures the spread of the action's consequences across the community: how many people's happiness is affected by the action?

In *Principles of Morals and Legislation*, and in unpublished manuscript fragments that were discovered in the late nineteenth century at University College in London by the French historian and philosopher Élie Halévy, Bentham described what we would nowadays call an *algorithm*, which could be used to compute total felicity. Intensities of pleasure or pain are to be measured as multiples of the faintest sensation that one can just make out, sometimes called a *hedon*, after *hedonism*, or a *dolor* (after the Spanish word for "pain"). Duration is measured in units of time, such as minutes. Probability is expressed as a fraction between zero and 1, with 1 denoting certainty. Propinquity is also a fraction between zero and 1, where 1 denotes immediate occurrence and zero denotes the end of one's life. Now the computations start. The number of hedons is multiplied by the duration in minutes, and the product is multiplied by the two fractions that denote probability and imminence. For dolors, the same computation is performed, but the result is preceded by a minus sign. This is repeated for all subsequent pleasures and pains that arise as a consequence of the action (i.e., the action's fecundity and purity). The results are added to obtain one individual's total happiness or utility.

At this point of the algorithm, a problem arises. If everybody were homogenous, one could simply multiply the result just obtained by the number of people affected by the action. But while duration, probability, propinquity, fecundity, and purity are more or less objective traits, Bentham realized that not all individuals experience pleasure or pain with the same intensities.[22] Hence, the computations must be redone separately for each and every person, taking into account the number of hedons and dolors that are appropriate for each individual. The results should be added up, and *now* we have the decision tool: A government that must choose between different courses of action will adopt the one that gives the highest result.

The felicific calculus opens a Pandora's box. How does one compare an action whose effects are more intense but shorter-lasting than another? Are 5 hedons that last for 6 minutes equivalent to 6 hedons that last for 5 minutes (as it would in ordinary multiplication)? Do they compensate for 4 dolors that last for 7 minutes? Above all, are Peter's hedons equivalent to Mary's? If not, adding them together would be like mixing apples and oranges. For all these reasons, the felicific calculus was not meant as a practical procedure but rather just as a proof of concept.

Whatever the impracticality of the algorithm, happiness was, by Bentham's reckoning, the measure of all things. How about money? What was its role? Bentham brought up money, for example, in 1801 in *Principles of the Civil Code*. In Chapter 6 of Part I, several axioms link wealth and happiness. The first two are:

1. Each portion of wealth is connected with a corresponding portion of happiness.
2. Of two individuals, possessed of unequal fortunes, he who possesses the greatest wealth will possess the greatest happiness.

Axiom 1 establishes an equivalence between money and happiness. This may seem obvious at first, but it is not, of course—as the saying goes, money does not buy happiness. But again, although money does not buy happiness, it does help. Even a profoundly unhappy person will become just a wee bit happier after the receipt of a dollar. And each additional dollar increases happiness, Bentham maintained. He drives the point home in Axiom 2: The more money a person has, the happier he is. Of course, half a century later, Karl Marx would have none of that. He would call Bentham a "genius in the way of bourgeois stupidity".[23]

Notwithstanding Karl Marx's low opinion of Bentham and Bentham's misgivings about the Declaration of Independence, this document's insistence on the relentless and never-ending pursuit of happiness as a right of all men, including those who are already rich, seems to answer and justify a deep desire of human nature. The Virginia Declaration of Rights, which served as a model for the Declaration of Independence that would be drafted a few months later, was even more specific: It stated that everybody has certain inherent natural rights, but they were not limited to life, liberty, and the pursuit of happiness. Aligning themselves with Locke, the "Representatives of the good people of Virginia" had gone a step further than he had: Among the inherent natural rights of all men, they also counted, and noted in the first paragraph, "the means of acquiring and possessing property." When Thomas Jefferson drafted the Declaration, he omitted the reference to property, substituting "happiness" instead. By replacing mere property rights with a much broader concept, Jefferson extended rights also to those who do not hold, or may have no interest in holding, tangible property. Thus, it was left to every person to interpret happiness as he wished, be it as property or anything else. However, in no way did Jefferson belittle (or negate) the right to private property. The opposite is true, as he affirmed in a letter to the French émigré economist Pierre Samuel du Pont de Nemours: "I believe . . . that a right to property is founded in our natural wants, in the means with which we are endowed to satisfy these wants."[24]

It should be noted that according to the Declaration of Independence, every man has the right to pursue, but not to *attain*, happiness. The latter would indicate that there is an absolute level that should be reached, while the former means that happiness is an elusive objective, for which ever higher levels can be reached. No matter how happy or rich one is, one has the right to pursue additional happiness, which, by Bentham's Axiom 2, is tantamount to pursuing ever more money and wealth. Contrast this, for example, with the minimalist view taken by the United Nations. Its Universal Declaration of Human Rights affirms only that "everyone has the right to a standard of living adequate for the health and well-being of himself and of his family."[25] And the International Covenant on Economic, Social, and Cultural Rights calls only for "remuneration which provides all workers, as a minimum, with . . . a decent living for themselves and their families . . ."[26] That does leave open the possibility of being paid more than a strict minimum, but confers no right to a higher pay.

Let us leave aside the United Nations, who, after all, only strives to achieve the achievable rather than chase lofty ideals. Philosophers, antique and modern, who thought deeply about money and wealth reached the obvious, if unsurprising, conclusion, formulated by scholars from Aristippus and Epicurus to Locke, Bentham, Jefferson, and beyond: More money is better than less. So ingrained is this maxim in human nature that it seems almost tautological.

Mathematically, this means that in a graph on which money or wealth is plotted along the x-axis, and the corresponding utility on the y-axis, the line always rises. In even more mathematical terms, the first derivative of the function that describes utility as a function of wealth is always positive.

CHAPTER 3

...AT A DECREASING RATE

The subject of chapter 2—that more of any good (money being one of them) is better than less of it—seems obvious.[1] But there is more to the story than this tautological-sounding assertion. And, as with so many deep questions, the discussion began in ancient times, with the most influential Greek philosopher, Aristotle, weighing in.

With his deep insight into human nature and the behavior of men, Aristotle (figure 3.1), who is considered by many as the father of psychology, had a lot to say about wealth and how it is perceived. In the *Nicomachean Ethics*, for example—the famous work named after his father or his son, both of whom were called Nikomachos—the philosopher held forth on how one should live an honorable, righteous, and happy life. In *Book IV*, one of the characteristics that Aristotle deemed desirable for a virtuous man is generosity. A magnanimous man should give of his wealth to whom and as much as is proper. Among the charitable and socially beneficial acts that Aristotle mentions are the building of temples, offerings to the gods, endowment of choirs, equipping a battleship, or offering a feast for the city's inhabitants. The virtuous person should donate willingly and happily because the virtuous action itself is what makes a person happy. Whoever donates unwillingly, or for reasons other than ethical motivations, cannot be considered generous— for he prefers money to the noble deed.

On the other hand, one must not overdo it. One should neither donate blindly to the first comer, nor give to the undeserving, nor distribute at an inopportune time. Oh, and by the way, one must not neglect one's own wealth,

FIGURE 3.1: Aristotle.

Source: Wikimedia Commons; Ludovisi Collection

Aristotle cautions, because that would diminish one's ability to continue to contribute to the needy and to noble causes. Whoever squanders his wealth on philanthropic causes cannot be considered generous, but rather must be regarded as a spendthrift. The question that poses itself is: What is the proper amount that a person should donate to be considered noble and generous?

Here is where Aristotle introduced a novel idea. Instead of delving into numbers, he approaches the question from a different angle, proposing a hierarchy of philanthropic characters. First there are the magnanimous.

They distinguish themselves by gifts that are so substantial that only the rich can (and should) afford them. If a not overly wealthy person tried to emulate his more prosperous fellow citizen by donating above his means, he would face insolvency, thus putting at risk his ability to do more good in the future. Next come the generous who support noble causes for their own sake. They value money not for what it is, but for its ability to do good. Nearly at the bottom (but not quite) are the stingy, the closefisted, who haggle about prices, cut corners, and never stop complaining about high costs, even when they do organize a noble event. Thus, they rob a gallant deed of its beauty. Ranked even below the miserly are the boastful braggarts, who would, for example, treat a group of friends to so extravagant a dinner party that it could pass as a wedding feast. Their apparent generosity is only a pretense because they do not act out of kindness or a sense of civic responsibility—they simply want to show off their wealth to gain the adulation of their peers.

The fact that Aristotle never mentioned numbers in his taxonomy of benefaction is significant. He emphasized that the size of a contribution is a relative concept. For example, the expenditures of a Trierarch, a citizen who equips a whole galley for public service, are different from those of an Architheoros, who only funds a religious ceremony. The seemliness of one's contributions is measured not only according to the costs of the object or event, be it a battleship, a temple, or a feast for the citizenry, but, above all, according to the donor's economic situation. The correct action should be guided by the means that are at a donor's disposal. Hence, while it is fitting for a rich man to fund a temple, a sacrificial offering, or a sporting event, it would be unseemly for a less well off individual to do the same. The main point is that, according to Aristotle, magnanimity must not be assessed in absolute terms. "Generosity," he wrote, "must be measured in relation to a man's wealth. It does not depend on the size of the gift but on the donor's circumstances. A generous person gives in proportion to his wealth. It may well be that a smaller gift signifies greater generosity, if it derives from smaller wealth."[2]

The view that a sum of money should not be measured in absolute terms, but rather ought to be considered in proportion to the person's wealth, is spelled out even more clearly in another of Aristotle's most important works. *Politika* is a treatise on political philosophy, and because economics is an integral part of that field, it plays a prominent role in the text. There are three sources of contentment, Aristotle claimed in *Book VII*:[3] external goods (i.e., earthly belongings); goods of the body (namely, physical well-being); and spiritual goods (e.g., courage, temperance, and strength of character). Of the spiritual goods, and presumably also the goods of the body,[4] it is always better to have more than less, with no limit on the maximum. Most interesting is what Aristotle has to say about external goods, the ones we equate with wealth, like gold, real estate, and cattle: "External goods have a limit, like any other instrument, and all things useful are of such a nature that where there is too much of them, they must either do harm, or at any rate be of no use to their possessors."[5]

Does that not ring a bell? Remember what happens when one consumes too many cookies or too much ice cream. According to Aristotle, satiation, or even nausea, is not the only reason why utility would increase at a decreasing rate, if it increases at all. Many external goods—such as fields, oxen, and plows—are simply tools that are meant to augment one's wealth. Beyond a certain point, there will be so much of them that concern for their maintenance will surpass their usefulness. Aristotle states clearly that, for whatever reason, beyond a certain point, additional wealth affords no additional utility.

After this detour to Greek philosophy, we revisit Jeremy Bentham. He had much more to say about the utility of wealth than I recounted in the previous chapter. In particular, in his attempts to create a universal code of law, he stipulated not only the two axioms from the previous chapter, but several additional ones. Whereas Axiom 1 established an equivalence between happiness and money, and Axiom 2 affirmed that more money is better than less, Axioms 3 and 4 in the *Pannomial Fragments* describe how utility rises with increasing wealth. Bentham published *Principles of the Civil Code* in 1801, and he wrote the *Pannomial Fragments* at various times throughout his life, with the last notes dating from about 1831. While it is not known whether he was familiar with, or even aware of, Daniel Bernoulli's solution to the St. Petersburg Paradox of 1733, Bentham actually proposed something identical, minus the mathematical trimmings. He realized that human nature is such that people place less utility on each additional unit of wealth than on the previous unit. He considered this insight of such fundamental importance that it needed to be formulated as a number of axioms. In *Pannomial Fragments*, he stated axioms 3 and 4 as follows:

> Axiom 3: But the quantity of happiness will not go on increasing in anything near the same proportion as the quantity of wealth: ten thousand times the quantity of wealth will not bring with it ten thousand times the quantity of happiness. It will even be [a] matter of doubt, whether ten thousand times the wealth will in general bring with it twice the happiness.

and

> Axiom 4: The effect of wealth in the production of happiness goes on diminishing, as the quantity by which the wealth of one man exceeds that of another goes on increasing: in other words, the quantity of happiness produced by a particle of wealth (each particle being of the same magnitude) will be less and less at every particle; the second will produce less than the first, the third than the second, and so on.[6]

As axiomatic systems go, these two statements are a bit of overkill. Systems of axioms, such as Euclid's, should be short, concise, and precise, containing no more than the bare minimum of information needed to explain everything of interest within that system, while disallowing any contradictions. Importantly, axioms should not be repetitive because different wordings may give rise to confusion and possible contradictions. Adding two axioms to express the same idea, as Bentham did with Axioms 3 and 4, is an absolute no-no. However, the insight that utility increases, albeit at a decreasing rate, may have seemed to Bentham so surprising, and maybe even counterintuitive, that he found it prudent to drive home the point with more than just a single axiom, without worrying about conciseness.

Bentham followed up with some practical consequences of that insight. Because "by a particle of wealth, if added to the wealth of him who has least, more happiness will be produced than if added to the wealth of him who has most," it is advantageous, from an aggregate happiness point of view, to take from the rich to give to the poor. Bentham formulated this insight in garbled language, as follows: "Income of the richer loser, £100,000 a year . . . income of the less rich gainer, £10 a year . . . wealth lost to the richer, gained by the less rich, £1 a year . . . on the sum of happiness the effect will be on the side of gain. More happiness will be gained by the less rich gainer, than lost by the more rich loser. A particle of wealth, if added to the wealth of him who has least, more happiness will be produced than if added to the wealth of him who has most."[7] Well . . . we see why the *Fragments* are called fragments.

In standard English, this would read as follows: If 1 pound is taken from a rich man and given to a poor man, the sum total of happiness increases. This conclusion, which follows immediately from the axiom that utility or happiness increases at a diminishing rate, is the origin of progressive taxation.

Bentham cited another consequence of this axiom. Let us say that a legislator has 10,000 pounds at his disposal for distribution among his constituents. Let us further assume that a poor person values an additional 1 pound half as much as rich people value the addition of 10,000 pounds. If the legislator distributed 1 pound each to 10,000 poor people, he would create 5,000 times as much happiness as he would have produced had he given all 10,000 pounds to one rich man. It is quite surprising how much happiness a legislator can produce with the stroke of a pen, but the assumption that Bentham posited is probably too strong. A more reasonable assumption, transposed into a contemporary situation, would say that a middle-class person values an additional dollar 100 times as much as does a billionaire. Then the distribution of $10,000 among 10,000 middle-class people would still produce 100 times more happiness than giving the whole sum to the billionaire.

A third example again concerns the transfer of a unit of currency from one person to another. As an incidental benefit, Bentham applied this example to illustrate the evils of gambling. In the patchy style of the *Pannomial Fragments*, the relevant comment reads as follows:

> "Income of the richer, say £100,000 a year . . . income of the less rich, say £99,999 a year: wealth taken from the first, and transferred to the less rich, £1 a-year . . . on the sum of happiness the effect will be on the side of loss . . . more happiness will be lost by the richer than gained by the less rich. Hence one cause of the preponderance produced on the side of evil by the practice called gaming."[8]

Unfortunately, here Bentham erred—though only on the details, not the principle. What he said is that when the 100,000-pound person loses 1 pound to the 99,999-pound person, the loser gives up more happiness than the winner gains. But actually, after the gamble has been made, the situation will be exactly the same as it was before, only with reversed roles. The formerly poorer person will now earn 100,000 pounds, while the formerly richer man will now earn 99,999 pounds. Hence, aggregate happiness is exactly as it was at the outset. Had Bentham said that the richer person earns 100,001 pounds and loses 1 pound to his gambling partner with 99,999 pounds, they would end up with 100,000 pounds each. Only in this case will total happiness actually increase: The poorer person has gained more happiness through the addition of 1 pound than the richer person has lost by giving up 1 pound. So, this does not make a good argument against gambling. What Bentham may have wanted to say is that if the poorer person *loses* 1 pound to the richer person, then, and only then, will total happiness have diminished. (The point can also be made, and more easily, if two people of equal wealth gamble for 1 pound.) It is quite gratifying to realize that even a philosopher of Bentham's stature may err sometimes.

This led Bentham to another conclusion. Because total happiness increases when money is transferred from a wealthy person to a less wealthy one, it follows that the process should continue until everybody owns the same amount. "The more nearly the actual proportion approaches to equality, the greater will be the total mass of happiness," he stated in *Principles of the Civil Code*. In the *Pannomial Fragments*, he even advises that decision-makers should actively promote such redistribution of wealth: "On the supposition of a new constitution coming to be established, with the greatest happiness of the greatest number for its end in view, sufficient reason would have place for taking the matter of wealth from the richest and transferring it to the less rich, till the fortunes of all were reduced to an equality."[9] Surely Karl Marx would have agreed with this; apparently, he overlooked that part of Bentham's writings when he called him a "genius of bourgeois stupidity."[10]

In any case, all these declarations—take from the rich, give to the poor, gambling is out, equality is in—follow straight from the axiom that each additional unit of wealth produces less happiness than the previous one.[11] However, they depend on the assumption that the utility, or happiness, that different people ascribe to money can be compared, and indeed is equal. Unfortunately, because utility cannot be compared directly among individuals, the declarations, though possibly true, would still need to be proved to satisfy the rigorous standards of mathematical economists. We will return to that point in later chapters.

By stating as an axiom the fact that utility increases at a diminishing rate, Bentham simply noted a basic principle of human behavior that Bernoulli had discovered seventy years earlier . . . to say nothing of Aristotle, who had surmised as much two millennia before that. The difference between Bernoulli's and Bentham's ways of expressing this principle was that Bentham, the political scientist, expressed it in qualitative terms, while Bernoulli, the mathematician, gave it a mathematical guise. The thinkers who followed needed to find a way to ascribe precise numbers to the elusive utilities.

The first to pick up where Bernoulli had left off was the Frenchman Pierre-Simon Laplace (figure 3.2). Born in 1749, one year after Bentham, Laplace was among the foremost mathematicians, physicists, and astronomers of the eighteenth century, as the Bernoulli family had been in the preceding century. As such, he was also one of the first scientists to raise probability theory onto a sounder theoretical foundation from the rather impressionistic basis where it had lingered for the previous century and a half.[12]

In his *Essai philosophique sur les probabilités* (Philosophical essay on probability), Laplace stated, as Aristotle, Bernoulli, Bentham, and others had done before him, that utility increases at a decreasing rate: *"l'avantage moral qu'un bien nous procure n'est pas proportionnel à ce bien . . . il dépend de mille circonstances souvent très difficiles à définir, mais dont la plus générale et la plus importante est celle de la fortune."* (This translates as, "The utility that a good brings is not proportional to that good . . . it depends on a thousand circumstances that are sometimes difficult to define, the most general and most important of which is wealth.") In fact, he continued that "it is obvious that one Franc has much more value to a person who owns no more than a hundred than to a millionaire."[13]

In his magisterial *Théorie analytique des probabilités* (Analytical theory of probability), first printed in 1812 and republished no less than five times during the next thirteen years, he devoted a whole chapter to *l'espérance morale* (i.e., to expected utility). Indeed, ninety-nine years after Nikolaus Bernoulli had first raised the issue, and eighty-one years after his cousin Daniel had

FIGURE 3.2: **Pierre-Simon Laplace.**

Fig. 5. — Laplace.

Source: Wikimedia Commons

proposed the solution, Laplace embraced the mathematical side of utility theory to the full, crediting Daniel Bernoulli with its discovery. In chapter 10, he took the reader through the painstaking calculations that he performed. He computed that a person with an initial wealth of 100 francs, whose utility for money is determined by the logarithmic function, would be willing to pay no more than 7.89 francs to participate in the St. Petersburg game. And if he had started with a wealth of 200 francs, he would have been willing to increase his stake by no more than 89 centimes, to 8.78 francs.

Laplace then went on to apply this newfound method to a more serious challenge than the problem of how much money to stake in a coin-tossing game. The subject he turned to was a hot topic at the time—old-age pensions. Until the French Revolution, the guilds—to which all craftsmen had to belong in order to ply their trade—had taken care of their members in old age. But in 1791, in an attempt to liberalize commerce and industry and create free competition, the government decreed the abolishment of guilds. Henceforth, everybody was allowed to practice any profession he wanted. An undesirable side effect of this decree, however, was that the bond that had hitherto existed between active and retired members of the guilds was broken, and the old were left to fend for themselves. Private associations, the *sociétés de secours mutuels* (societies for mutual support), which craftsmen could join voluntarily, jumped in to fill the gap. Dues-paying members of these associations were assured that they would be taken care of financially in old age and in case of illness. But the associations were rather loosely organized. Hence, their continued existence in the far future, when their financial soundness would really matter to the members, was a matter of profound uncertainty. Only in 1835 was a law enacted that would regulate the affairs of the *sociétés*.

When *Théorie analytique des probabilités* first appeared in 1812, the subject of old-age pensions was in limbo, and life insurance was a topic of great interest. The specific question Laplace posed, as an example of the theory's usefulness, was: Would it be advantageous, utilitywise, for a couple to take out a life-insurance policy jointly, or should they each buy individual policies? By applying arcane mathematical tools to a question that concerned everyone at the time, Laplace was sure to gain his readers' attention. He posed the problem as follows: Under the joint policy, each spouse would receive half the pension, so long as both of them were alive, and the full amount after the first partner died. With individual policies, each pension would terminate upon the death of the insured. One condition that Laplace placed on the pension payments—a condition on which the life insurance company would insist—was that the discounted expected payouts under both options must be identical. (Hence, the yearly pensions that were to be paid out would be different.) Using mortality tables, interest rates, and discount factors, and assuming that both partners have a logarithmic utility for wealth, Laplace arrived at the conclusion that it was advantageous for a couple to insure themselves jointly.[14]

Having thus disposed of the dry mathematics, Laplace turned surprisingly starry-eyed. Venturing beyond pure calculations, he wrote that at the death of a spouse, the surviving partner, who is presumably at an advanced age, will benefit from higher pension payments from the joint insurance policy, precisely at a time when he or she will need it most. Given the affection that prevails between the partners—and here this sober mathematician turned decidedly romantic—a person lying on her or his deathbed would wish for nothing more than the continued well-being of the surviving partner and family.

Because it is not often that financial institutions, morality, and sweet desires are mentioned in a single sentence—and by a mathematician, mind you—I want to quote Laplace in full: *"Les établissements dans lesquels on peut ainsi placer ses capitaux et, par un léger sacrifice de son revenu, assurer l'existence de sa famille pour un temps où l'on doit craindre de ne plus suffire à ses besoins, sont donc très avantageux aux moeurs, en favorisant les plus doux penchants de la nature."*[15] (Or, "Financial institutions in which one can place one's capital, thus assuring—with a slight sacrifice of income—one's family's continued existence during a time when one must fear of not being able to satisfy its needs, are very advantageous for morality, thus favoring the sweetest desires of human nature.") With financial institutions nurturing morality and sweetness, Laplace advised governments to encourage them because—given the hopes far into the future that they embody—they can prosper only if all concerns about their continued existence are dispelled.

Having moved from Bernoulli, the mathematician, to Bentham, the political scientist, to Laplace, the probabilist, it is time for the baton to pass to explorers of the human body and mind. Two Germans now take center stage: the medical doctor Ernst Heinrich Weber (figure 3.3) and the psychologist Gustav Theodor Fechner in Leipzig.

Born in 1795 to a preacher and eventual professor of dogmatics, Michael Weber, Ernst Heinrich was the third of thirteen children, of whom seven survived. A precocious boy, who made his name at school not only as an outstanding student but also as a wrestler, he began his medical studies at age sixteen at the university in Wittenberg. It was in this town that the reformer Martin Luther had proclaimed his ninety-five theses three centuries earlier, with which

FIGURE 3.3: Ernst Heinrich Weber.

Source: Wikimedia Commons

he challenged the Catholic Church. Wittenberg was also the hometown of the physicist Ernst Chladni, whose Chladni figures—formed by the vibrations that sound produces in metal plates that are covered with a thin coat of sand— would become important not only for their surprising beauty but also for their influence on the design of violins, guitars, and cellos. Chladni was a frequent guest at the Weber house and formed a friendship with Ernst Heinrich that would last until his death.

In 1814, only three years into Ernst Heinrich's studies, the turmoil of war forced a change of plans. The Prussian army recaptured Wittenberg, which had been under the tutelage of the king of Saxonia, a protegé of Napoleon. The fighting brought great destruction to the city; in fact, the Weber family had to flee their burning house after a hit by the French artillery. The damage to Wittenberg was so great that the university had to relocate to a small town nearby. Eventually, the elder Weber picked up his family and moved to Halle.

Armed with a recommendation by Chladni, Weber seized the opportunity to transfer to the University of Leipzig, where he delved into the study of human and animal anatomy. He wanted to understand the organs' functions and their interplay by relating the structure of the organs of certain animals to their way of life. Using a hunting rifle and a fishing rod, he collected many of the (nonhuman) specimens for his research himself. In 1815, he obtained his medical degree with a thesis on comparative anatomy.

His subsequent career advanced swiftly. While practicing as a medical doctor, he continued his research in anatomy, notably with groundbreaking work on the hearing organ of fish; and he was named a full professor of anatomy in 1821. Thus, at the age of only twenty-six, he had already reached the pinnacle of German academia. According to tradition, now was the time to build a homestead, and Weber married Friederike Schmidt, the sister of a friend from his youth. He would share more than fifty years of marital bliss with her.

Despite the expenses that he incurred with the establishment of his new home, Weber spent a significant amount of money on the purchase of a bowl of mercury, then considered a medicine against glandular disorders, that he would use for his anatomical explorations. Even though the originally planned experiments would be only partially successful, the purchase turned out to be a boon for science. One day, when cleaning the liquid of dust and impurities, Weber observed movements on the viscous surface that reminded him of Chladni's figures. This led him and his younger brother Wilhelm, whom he coopted while the latter was still attending high school, to study the emergence and behavior of waves all the way down to the oscillations of molecules. The ensuing experiments, conducted at their father's house because the laboratory at the university was too small, took four years—during which, week after week, the elder brother undertook the twenty-mile journey from Leipzig to Halle on foot because there were no bus lines and certainly no trains running between the two cities. The brothers' joint work culminated in the volume *Wellenlehre auf Experimente gegründet* (Textbook on waves, based on experiments), that they dedicated to their friend, Ernst Chladni.

So thorough was the book that the two brothers soon became household names in physics.

One of Ernst Heinrich Weber's most important works, however, was *Handbuch der allgemeinen Anatomie des menschlichen Körpers* (Manual of general anatomy of the human body). If the number of editions that follow the first publication of a textbook is a measure of its importance, Weber's *Handbuch* was unsuccessful—there were no reprints. What had happened was that a pirated edition appeared on the market, which made any reprints superfluous. This did not diminish the original work's influence on later writers, however. Subsequently, Weber became involved in local politics, balancing the needs of the University of Leipzig with those of the city as a whole.

Convinced that the teachings about the organs' form and function must be linked, Weber lectured on physiology, without reimbursement, in addition to anatomy. In the dissection exercises with his students, he was very ably assisted by his brother Eduard Friedrich, later a professor of anatomy. (Wilhelm, the third brother, who would become a professor of physics at the University of Göttingen, would write a book with Eduard Friedrich on the mechanics of human walking.)

The general conditions in the university's anatomical institute were dismal. Weber's description speaks for itself:

> [O]ne floor and an attic in a small building, with one large unheatable hall used for lectures, that also holds part of the anatomical collection; another heatable room that is used during winters, both for lectures and for exercises in dissection . . . the collection of specimen is dispersed throughout various chambers . . . the institute lacks running water and there is no drainage . . . foul-smelling liquids must be carried down the stairs in canisters and disposed of outside the building, which makes anatomical work difficult and renders the institute burdensome to the neighbors . . . the narrow room in which lectures are given, corpses are prepared for dissection, and exercises are held, has only three windows . . . only about half of the students who participate in the exercises have work places that are sufficiently lit . . . during lectures, many students must stand on tables and still cannot see anything.[16]

The shortcomings at the institute were not only of an architectural nature. At times, a lack of corpses threatened instruction in anatomy, and lectures had to be suspended because there were no cadavers on which to demonstrate. Generally, the University of Leipzig relied on the corpses of suicides, deceased prison inmates, paupers, and accident victims with no next of kin. But it had to compete for them with the University of Dresden, about seventy miles farther east. Some relief came when the faculty of medicine at Dresden closed its doors, but the threat of a cadaver shortage loomed anew when the burghers of a nearby town were about to decide that even suicides and the truly poor deserved proper burials. Weber protested vehemently against this idea and convinced the ministry in charge that such a decision would cause irreparable harm to the University of Leipzig.

Upon reaching the age of seventy-six, Weber asked to retire. Although still capable of continuing his duties at the university, he wanted to make way for a successor who would fashion the institute according to his own wishes. During his long and illustrious career, he had never ceased to pursue new achievements as a scientist, even while serving as a husband and father, the people's representative in the local parliament, an administrator at the university, and a teacher at his institute.

The reader may be wondering what importance this professor of anatomy has to our narrative. Well, Ernst Heinrich Weber's fields of interest included the human senses, and it is this area that has great consequence to our story. As it turns out, Weber's experimentally derived research lent credence to Bernoulli's hypothesized utility of wealth!

The professor of anatomy and physiology had always been curious about the sensory organs, ever since his early investigations into the hearing of fish. In 1846, when he was already past fifty, he published *Der Tastsinn und das Gemeingefühl*, a book on the tactile and other senses that is nowadays probably the best remembered of Weber's works. The title cannot be translated easily because the word *Gemeingefühl* has no counterpart in English. In fact, the term may have been coined by Weber himself, specifically for this book. He wanted to convey a sense of body awareness (i.e., the sum of all sensations simultaneously), in contrast to the perception of individual sensations. Weber himself suggested the term *coenaesthesis* as a translation (from the Greek *kainos*, common, and *aesthesis*, sensation), but a more successful translation of the book's title may be "The sense of touch and the common sensibility." And as if the book's title were not complicated enough, its reader-friendliness left a lot to be desired. Sentences of more than eighty words were no rarity, and when one considers that the German language allows the concatenation of several expressions into a single word, one realizes how difficult the reading was. Actually, the book *Tastsinn* had a precursor, which Weber published in Latin in 1834. *De subtilitate tactus* (On the precision of the sense of touch) also dealt with tactile senses.

The questions that were explored in these two books included how the nervous system of the skin leads to the perception of space, and how humans perceive differences in weight or temperature through their muscles, nerves, and skin. In one famous experiment, Weber placed the two ends of a compass on the skin of human subjects—he had first filed off the pointed ends, so as not to prick the testees—and asked them whether they felt, without looking, a single impression, or two separate ones. In the first case, the compass's legs were so close together that the impressions fused into one. He then increased the distance between the two legs of the compass until the subjects could distinguish two separate impressions on their skin. It turned out that different parts of the body are vastly different in their response to stimulation. The fingertip and the tip of the tongue are much more able to distinguish close-by impressions than the thigh or the upper arm, with the forehead and the back of one's hand somewhere in the middle. While the tongue and lips can distinguish impressions that are separated by only a couple of millimeters, the skin on a thigh would require a distance of about 5 centimeters to tell them apart. Weber explained that this stemmed from the differing densities of nerve endings on the skin in these parts of the body. He was right: Recently, neuroscientists have found that larger parts of the brain are dedicated to those parts of the skin that have higher densities of *mechanoreceptors*, as such nerve endings are now called. Other experiments that Weber performed dealt with how human beings perceive different weights.

It is precisely because of these experiments that Ernst Heinrich Weber is of such importance for our narrative. Specifically, the implications of these experiments give credence to what Daniel Bernoulli had surmised a century earlier, albeit on a theoretical level and in a completely different context. Weber performed many experiments, putting weights on different parts of the body to see how fingers, the forehead, and the lower arm reacted; having subjects rest their hands on a table or hold them in the air while loading them with weights; using hot and cold weights; and running comparisons simultaneously or consecutively, with elapsed time spans between 15 and 100 seconds. He also devised a method to ascertain whether the muscles are responsible for the perception of weight, or the nerves. By putting weights on the last two phalanges of the fingers, while the hand rests flat on a table, Weber ensured that no muscles were involved. On the other hand, by having the subject grasp a piece of cloth with one hand and hold it out in front, while he placed weights into it, he ensured that only muscles were involved.

Weber found that when using muscle power alone, most people could tell 80 ounces (about 2.4 kilograms) from 78 ounces.[17] Using only the tactile sense—fingers flat on the table—subjects were able to distinguish 14.5 ounces (a little less than a half-kilogram) from 15 ounces. This led him to investigate what he called the *"kleinste Verschiedenheit"*—the smallest, or just noticeable difference (JND). The chapter that is the most relevant for our story carried the title "On the smallest differences in weight, that we can differentiate with the tactile sense, lengths, that we can differentiate with the visual sense, and tones, that we can differentiate with the auditory sense."[18] He did not stop there. Weber also investigated people's ability to note differences in a light's brightness, the roughness of a surface, and the hardness and temperatures of materials.[19] Weber also mentioned smells, even though there was no way to quantify different smells along a single dimension.

A human's ability to distinguish between weight differences can be fine or coarse, Weber found, depending on which part of the body is employed. A finger, the more sensitive of the body parts that he studied, could distinguish weight differences of 3 to 4 percent. When weights were placed on their foreheads, subjects could note differences of about 6.5 percent. But when weights were placed on the middle of the forearm, test subjects were unable to distinguish any difference in weight of less than about 10 percent. When perceiving lengths of lines on a piece of paper, the crucial variable was how much time elapsed between the trials. If they occurred within three seconds of each other, subjects could note differences in length of about 2.5 percent. With a gap of thirty seconds, the length differences had to be at least 5 percent. If seventy seconds or more elapsed, even a difference in length of 10 percent was no longer noticeable.

However, we are not concerned with the actual numerical results or the intricacies of the innumerable experiments—whether different weights are placed simultaneously on both hands or consecutively on the same hand, whether a metal rod seems colder than a wooden rod of equal temperature, and so on. What is of much greater interest to us is that Weber realized that the JND must not be expressed in grams or ounces, in millimeters or fractions of inches, or in degrees Centigrade or Fahrenheit.[20] It is not the absolute value of the additional weight or the additional length or the increase in pitch that is of importance, but the *relative* change, in proportion to what existed at the outset. The heavier, the longer, the higher the initial values, the more the weight, the line, or the pitch must be increased for someone to detect JNDs. Hence, the only correct way to describe JNDs is to express them as percentages.

The fact that proportions—not exact, numerical measurements—are required to judge differences, was deemed an "extremely interesting psychological phenomenon" by Weber. "I have shown that success in the judging of weights is the same, irrespective of whether one uses ounces or lots," he wrote. "What is relevant is not the number of grams that form the surplus weight but whether the surplus weight is one thirtieth or one fiftieth of the load that is compared to another weight. The same holds true for comparisons of lengths of two lines and pitches of two tones."[21]

Now, recall that a homeless person would probably thank you profusely for a dollar, while a billionaire would require maybe a hundred thousand times as much to notice a difference in his wealth. Weber gave experimental proof, albeit in a completely different context, that Daniel Bernoulli (and, before him, Gabriel Cramer) had been on the right track. Hence, we may stipulate that the perception of additional wealth, like the perception of additional weight, depends on what one owns at the outset.

Weber's ideas were eventually taken up at the University of Leipzig by a former student and later colleague of his, Gustav Theodor Fechner (figure 3.4). Six years younger than Weber, Fechner grew up in poor circumstances. His father, a pastor, died when the boy was only five years old. His mother raised Gustav and his four siblings on her own, with some help by her brother, also a pastor. Fechner, who excelled in high school, decided on a medical career and first attended academic studies in Dresden. After one semester, he transferred to Leipzig, where he would remain, never to leave again until his death seventy years later.

Fechner did not enjoy his studies. Only the lectures on algebra and Weber's course on physiology caught his attention. It is a testament to the deplorable state of nineteenth-century medical education in Germany that he managed to pass his exams in pathology and therapy without having attended any lectures or labs, solely by studying the subject matter from books. He knew then that he was not destined to be a medical doctor. Always low on funds, he augmented his meager income with private lessons and literary work, while also writing and translating textbooks on chemistry and physics, lecturing at the university, and pursuing his own research in electrodynamics. Several results of his experiments, especially his inquiry into Ohm's law of electricity, were of some importance to nineteenth-century physics, and in 1831, he was promoted to a professorship of physics.

While teaching at the university, Fechner authored the three-volume textbook on experimental physics *Repertorium der Experimentalphysik* and edited the *Pharmaceutisches Centralblatt*, a biweekly journal that kept pharmacologists and pharmacists abreast of research results in Germany, France, England, Italy, and the Netherlands. These publications were meant for a specialized readership, however, so they did not generate great financial benefits. In the hope that books for a general public would bring him more money, he set about editing an eight-volume encyclopedia, *Das Hauslexikon*, that was published by the company *Breitkopf & Härtel* between 1834 and 1838. With Herculean effort, he himself composed about one-third of the entries. Unfortunately, his hopes remained frustrated. Too modest to beat the drum for their work, he and his publisher relied on word-of-mouth advertising, with foreseeable consequences: The encyclopedia remained a niche product.

But the workload had become extremely intense—he never stopped publishing scientific articles, as well as works of fiction, and after years of unrelenting labor, the stress finally caught up with him. Already afflicted with an eye disease due to his experiments on vision, which had included peering at the sun through colored glasses, he suffered a nervous breakdown. Afraid of going blind, he sank into a deep depression, refused to eat, and shunned all social contact. It was mostly due to the loving care of his wife Clara that Fechner eventually recovered from the depths of despair.

When in good health, he surrounded himself socially with circles of friends, at first young men of dubious renown, but later intellectuals like himself,

FIGURE 3.4: Gustav Theodor Fechner.

Source: Wikimedia Commons

who met frequently to discuss philosophy, science, and matters of the day. Fechner looked nothing like the life of the party: Balding, with thin lips drawn tight, what hair he had falling unkempt to his shoulders, the corners of his mouth drooping downward, and wire-rimmed glasses on his nose, he seemed hardly the epitome of entertaining collegiality. But appearances are deceptive, because the sour-looking professor had a very well-developed sense of humor. Proof of this can be found in several booklets that he published under the pseudonym "Dr. Mises." In these writings, ostensibly scientific treatises, humor and a biting sense of irony shone through. For all that, the booklets were more than a means of making extra money, or a hobby with which to entertain his friends. They revealed a rather surprising worldview.

Amazingly, this student of medicine and physics subscribed to a philosophical school for a while that, although popular among certain circles in Germany in the eighteenth and nineteenth centuries, had already been thoroughly discredited among most scientists of his time. How he came to believe in these controversial ideas, he described thus: "Through my medical studies I had become a complete atheist, alienated from religious ideas," he remarked. "I saw the world only as a mechanical assembly."[22] But then he happened across the *Naturphilosophie* of the German naturalist Lorenz Oken, a romantic and speculative philosophy of nature proposing that life was everywhere and permeated everything. In Oken's and like-minded people's view, the world and all its parts are alive and conscious. Not only humans, but also plants and animals, and even planets and the cosmos itself have inner lives and possess souls.

Such a grand, unifying view of the world appealed to Fechner, but it contrasted sharply with the prevalent mechanistic view of the world. This is probably why he concealed his beliefs, referring to them only cryptically in his humorous writings. But as his readings progressed, he became more and more disillusioned. Would Oken's speculations have allowed the discovery of the laws of science, he asked himself. He had to admit that the answer was no; Oken's approach was not the way to gain knowledge in science. And that eventually put an end to his dalliance with *Naturphilosophie*.

But his interest in the relationship between a human being's body and soul never waned, and when he learned, while still a student, of the experiments that the distinguished physiologist Ernst Heinrich Weber was performing at his university, he was immediately drawn to them. One of the reasons he found Weber's investigations so appealing was the notion that a quantitative relationship could govern how sensations were perceived, thus establishing a connection between body and mind. The existence of a mathematical link between bodily sensations and their admission into consciousness made the body-mind problem amenable to scientific study. The publication of Fechner's *Elemente der Psychophysik* (Elements of psychophysics), printed and distributed in 1860 by his friend Hermann Härtel, of Breitkopf & Härtel, signaled that psychology had matured into a science.[23] From then on, psychology could no longer be considered simply a collection of philosophical speculations.

In the preface to this groundbreaking book, Fechner defined psychophysics as the "exact theory of the relationship between body and soul." The term *exact* indicates, Fechner explains, that the theory cannot simply consist of philosophical "*aperçus*" that may seem plausible to a thinker sitting at his desk; rather, it must build upon actual experiments to obtain the necessary measurements. As a prime example, he points to the experiments conducted by Weber, whom he calls the "father of psychophysics."[24] Equally important, however, Fechner continues, are the mathematical underpinnings of the theory.

Fechner passes review on Weber's, other scientists', and his own experiments on the relationship between the physical and perceived intensities of the brightness of light, lengths of lines, weights of loads, loudness of sound, pitches of tones, differences of temperatures, and other stimuli. But while Weber's law pertained to the JND in stimuli, Fechner's research provided an actual model across a whole range of intensities. Weber had recognized that, say, at a weight of 20 ounces, the JND is 1 ounce, and at 40 ounces, the JND is 2 ounces:

$$JND = 1/20 = 2/40 = \ldots = constant.$$

Fechner cast Weber's observations into a mathematical model. Denoting the JND of weight-perception as $dP(W)$, weight itself as Q, and additional weight as dQ, he got

$$dP(W) = k \cdot dQ/Q.$$

Integrating both sides, he obtained the following equation for the perception of weight:

$$P(Q) = k \cdot logarithm(Q) + constant.$$

Based on their experiments and the model, Weber and Fechner established that the perceived intensity of a stimulus is proportional to the logarithm of its physical intensity. Compare this with Bernoulli's formulation for the additional utility of wealth, $dU(W)$ (see Chapter 1),

$$dU(W) = c \times dW/W,$$

and hence the utility of wealth,

$$U(W) = c \cdot logarithm(W) + constant.$$

Thus, Fechner did for sensory stimuli what Daniel Bernoulli had done for the utility of wealth 130 years earlier. Fechner was quite aware of Bernoulli's previous work in this area. In fact, it is not clear whether he developed the model and then discovered Bernoulli, or vice versa. In any case, in his *Elemente der Psychophysik*, he credits Bernoulli and Laplace for having formulated the equation that models the observations. Maybe Weber and Fechner would have liked to ascertain that their law also holds for the perception of wealth; alas, without millionaire subjects at hand, they could not perform meaningful experiments.[25]

To end this chapter, recall that in chapter 2, I remarked that mathematically, on a graph in which wealth is plotted along the *x*-axis and the corresponding utility on the *y*-axis, the line always rises. Hence, the first derivative of the function that describes utility as a function of wealth is always positive. This chapter established that the utility function bends downward, which means that the second derivative is always negative.

PART TWO

MATHEMATICS IS THE QUEEN OF THE SCIENCES...

CHAPTER 4

THE MARGINALIST TRIUMVIRATE

With the exception of the mathematicians Pierre-Simon Laplace and Siméon-Denis Poisson (1781–1840), and the mathematically inclined psychologist Gustav Fechner, few people took note of Daniel Bernoulli's work. For a century and a half, his groundbreaking paper remained rather obscure. This was not very surprising, as the St. Petersburg Paradox, which the paper sought to explain, seemed nothing more than a question for gamblers and a diversion for mathematicians. Nobody seemed to notice that the problem actually belonged to the realm of economics, which was treated at that time not by quantitative reasoning, but mostly by casual accounts, offhand observations, and anecdotal evidence. To employ mathematics to deal with an economic question was unheard of until the last quarter of the nineteenth century when, all of a sudden, three men from three different countries, writing in three different languages, in blissful ignorance of each other, had the identical idea: The notion that underlies all economic decisions is not money, but the utility it provides, and the treatment needs to be mathematical. The three were William Stanley Jevons in England, Léon Walras in Switzerland, and Carl Menger in Austria.

The fundamental question at the time was how the price of a commodity or product is determined. What factors determine its worth? It is surprising that this should be a problem because it is we, the people, who collectively determine prices. Does the amount of labor that goes into production determine the price of a good, as the classical economists Adam Smith and David Ricardo believed? No, because no labor is required to produce water, for example, but nevertheless, during a drought, it is more valuable than a diamond. Is scarcity the factor that determines price? Maybe, because nobody would pay much for diamonds if their glamour and beauty did not create a demand that outstripped the limited supply.

The answer which began to take shape during the late eighteenth century, was that it is utility, as Cramer and Bernoulli surmised, that determines what one is willing to pay for a good. It was a difficult idea to grasp. Labor would have seemed an evident factor, but utility? The situation brings to mind the difficulty that Isaac Newton faced when he proposed that gravity is what pulls an apple toward the ground. People could picture a donkey being pulled by a rope—but by gravity? So utility is an invisible rope that pulls everything together.

The best known of the triumvirate may be Jevons (figure 4.1), if only because he wrote in English, a language more prevalent than French or German. He was born in 1835 in Liverpool, to an iron merchant with strong interests in engineering, economics, and law. The father was said to have constructed, in 1815, the first boat made of iron that could sail on water, and also established himself as an author, albeit one of moderate renown, with a small book on law and a pamphlet on economics. Unfortunately, the family business, which had grown out of a nail-making enterprise in Staffordshire, eventually ran into financial trouble.

The difficulties did not occur only on his father's side. Jevons's mother died when the boy was ten; he was her ninth child. She had been a gifted woman, a poet, from an illustrious family in Liverpool. Her father, William Roscoe, was a lawyer and banker.[1] A social reformer, he advocated the abolition of the slave trade. Unfortunately, as honest and irreproachable as the Roscoe family was, both the grandfather and father were bankrupted—the former during a run on his bank in 1816, the latter during the financial crisis of 1848. These misfortunes, as well as the troubles that his own father's business had to face, sensitized Jevons at an early age to the vagaries of the markets and the economy.

While he was still a little boy, his mother read *Easy Lessons on Money Matters* to him, a book that the Archbishop of Dublin, Richard Whately—a former professor of political economy at Oxford—had written for schoolchildren. It was probably Jevons's first exposure to economics. In spite of his parents' penchant for law, economics, poetry, and history, he was educated at school in mathematics, biology, chemistry, and metallurgy. At age fifteen, he

FIGURE 4.1: William Stanley Jevons.

Source: Wikimedia Commons

began his studies at University College London (UCL), but his first attempt at an academic education did not last long. Financial problems at home forced him to seek gainful employment, and at his father's urging, he cut his studies short to accept the post of assayer at the mint in Sydney, Australia. Sydney had just been awarded the right to mint sovereigns, gold coins that consisted of 22 carat gold and a few parts silver and copper. It was the assayer's job to ensure that the coins contained the proper amounts of each element. The position was offered to Jevons through a chemistry lecturer at UCL, and the student undertook the three-month trip to Australia with great reluctance.

Nevertheless, the pay was good, and his work at the mint left him sufficient time to order his thoughts. After four years, Jevons had saved sufficient funds to return to England. His father had died while he was in faraway Australia, and he traveled home through South and North America and the West Indies. Back in London, he took up lodgings with his brothers and sisters. Now, at age twenty-four, he picked up at UCL where he had left off five years earlier, obtaining his bachelor's degree in eleven subjects within a year. This was followed with a master's degree in logic, philosophy, political economy and mathematics two years later, and he was awarded a gold medal for his academic success in logic, philosophy, and political economy.

His first publication, *A Serious Fall in the Value of Gold ascertained, and its Social effects set forth, with two Diagrams* (1863), did not fare very well. In a private letter, Jevons wrote:

> I have just received the bill for my pamphlet on Gold, the total cost of printing, advertising, etc., is £43, and the offset by sales only £10; only seventy-four copies seem to have been sold as yet, which is a singularly small number.[2]

In an entry into his journal, he admitted to self-doubts:

> Now, I suppose I am low because my essay on "Gold" is out, and as yet no one has said a word in its favour except my sister, who of course does it as a sister. What if all I do or can do were to be received so? In the first place, one might be led to doubt whether all one's convictions concerning oneself were not mere delusions. Secondly, one might at last learn that even the best productions may never be caught by the breath of popular approval and praise.[3]

In his next work on economics, *The Coal Question: An Inquiry concerning the Progress of the Nation and the Probable Exhaustion of our Coal Mines* (1865), he argued that the demand for coal would need to be increased in a geometrical progression for Great Britain to maintain its prosperity and industrial leadership. Though most brilliantly and engagingly written—as John Maynard Keynes, the famous economist, would later state—his arguments were unsound, the prophecies did not come true, and the text appeared overstrained and exaggerated. However, the booklet established Jevons as a serious writer on economics and statistics. One of his observations became known as the Jevons Paradox: When technology improves, the efficiency of energy increases and less of it is required to produce the same output; as a result, one would assume that the demand for energy should decrease. However, Jevons surmised, the opposite may be the case: Because the availability of energy (coal, in this case) becomes more abundant, its price decreases and the demand for it may actually rise.

Jevons need not have plagued himself with self-doubt, for by then, his reputation had become solid. In 1866, he was elected professor of logic, philosophy, and political economy at Owens College, which would become the University of Manchester. A few years later, he produced his first bestseller, *The Principles of Science*, a work that made important contributions to logic and statistics. In 1876, Jevons was named professor of political economy at his former alma mater, UCL.

Jevons first mentioned the notion of utility in a letter to his sickly brother Herbert, then at the Bank of New South Wales in New Zealand. The theory is entirely mathematical in principle, Jevons wrote of economic science. One can derive definitions, axioms and laws in almost as rigorous a fashion as with many geometrical problems. The central axiom, Jevons pointed out, was that "as the quantity of any commodity, for instance, plain food, which a man has to consume, increases, so the utility or benefit derived from the last portion used decreases in degree. The decrease of enjoyment between the beginning and end of a meal may be taken as an example."[4] He returned to the notion in *A General Mathematical Theory of Political Economy*, a paper that he presented to the British Association for the Advancement of Science in 1862. The most important law of the whole theory, he wrote, is that utility is "some generally diminishing function of the whole quantity of the object consumed."[5]

To Jevons's disappointment, the paper did not attract much attention, neither upon its first presentation to the British Association nor upon its publication in the *Journal of the Statistical Society* four years later. It was only when his magnum opus, *The Theory of Political Economy*, appeared in 1871 that his ideas began to have an impact. But from then on, there was no stopping economic science would never be the same again. Two editions of the book were published during his lifetime, a third was overseen by his widow, and a fourth by his son. Many more followed, and the book is still in print today, in several editions. It is considered one of the seminal works in economics.

At the beginning of chapter 1, Jevons set out the two principles that would guide him throughout the development of his theory. The first one was the notion of utility, of course. "Repeated reflection and inquiry have led me to the somewhat novel opinion, that value depends entirely upon utility."[6] The opinion was novel because the prevailing view at the time was that an object's value lay in the labor that was needed to produce it. "There are even those who distinctly assert that labor is the cause of value," he pointed out indignantly, taking aim at Karl Marx's *Das Kapital*, which had been published just four years earlier. But a careful analysis of how a man's utility changes as he possesses more or less of a commodity is all that is needed "to arrive at a satisfactory theory of exchange, of which the ordinary laws of supply and demand are a necessary consequence." He took pains to emphasize to his readers that he had not arrived at this conclusion in a hasty or ill-considered manner. In fact, he had thought about it for ten years and questioned the truth of his notions over and over again. He found no reason to doubt its substantial correctness.

The second principle concerned the general character of economic science, and it may have been even more revolutionary than the introduction of utility into the discourse. It constituted a break with the prevailing view of how economics should be studied. The established methods were narratives, plausibility arguments, anecdotes, and proofs by example. Jevons realized that to elevate economics to a rigorous science, on a par with physics and astronomy, a different approach was required: "It is clear that Economics, if it is to be a science at all, must be a mathematical science." And, to state the obvious, he added that "our science must be mathematical, simply because it deals with quantities."

Jevons was not referring to simple arithmetic—adding, subtracting, dividing, and multiplying—but to more advanced tools, namely, differential calculus. "We cannot have a true theory of Economics without its aid," he stated emphatically. To jostle economists out of their state of denial, Jevons even resorted to ridicule. They "might as well try to alter red light by calling it blue," he wrote sarcastically, heaping scorn on traditional economists' rejection of mathematical notation and arguments. He went even further. Whenever scientists, even in subjects like physics or astronomy, attempt to avoid mathematics in order to explain their theories to a general readership, the inadequacy of words and grammar to express complicated relations quickly becomes apparent, he wrote. For that reason, mathematical symbols "form a perfected system of language, adapted to the notions and relations which we need to express." Mathematical reasoning and mathematical symbols cannot be avoided; they are indispensable for the study of economics.

Jevons identified another reason why many of his contemporaries resisted the use of mathematical tools, apart from their general dislike for them. Economists often confused mathematical tools and exact science. He conceded that data in economics are not precise, but that this should not discourage economists from employing mathematical tools. After all, in astronomy, the positions of stars are known only approximately; geographers assume that the Earth is a smooth, homogenous spheroid; in statics, physical bodies are assumed to be perfectly inflexible; and the list goes on. All these supposed facts are only hypothetical approximations to the truth. "Had physicists waited until their data were perfectly precise before they brought in the aid of mathematics, we should have still been in the age of science which terminated at the time of Galileo," Jevons concludes.

After making his cases for mathematics and for utility, Jevons proceeded to the theory of exchange. As usual, he reformulated the problem as a question of how to maximize overall utility through the enjoyment of different amounts of various goods. Let us say that people must decide about the amounts of corn and beef that they would like to trade. Postulating that the utilities of additional corn and beef decrease as more of them are consumed, the question then becomes: How much beef will be exchanged for corn?

We assume that there are two traders: a beef-trader and a corn-trader, both of whom possess large amounts of their respective goods. Initially, the corn-trader, starved for beef, is willing to give up 20 pounds of corn for 1 pound of beef. Because his utility of beef decreases, he would exchange a second pound of beef for only 16 pounds of corn, a third pound for 14, and so on. For the beef-trader, who initially lacks corn, the situation is exactly the reverse. At first, he is willing to give up lots of beef for some corn. But the more corn he can obtain, the less utility he assigns to it, and the less beef he is willing to give in exchange for even more corn. The two traders will continue negotiating until they reach a point where their utilities no longer increased by exchanging goods. Let us say that they are willing to exchange 10 pounds of beef for 100 pounds of corn. This indicates not only the total amount of goods they want to trade but also the exchange rate. The result of their negotiations shows that corn provides one-tenth the utility of beef; hence the ratio of exchange is 10:1. Beyond that point, the utility to the corn-trader of an additional pound of beef drops to 9.5 pounds of corn, but the beef trader is not willing to exchange a pound of beef for less than 10.5 pounds of corn.

By comparing their respective utilities, the two dealers arrive at the law of supply and demand. They discover not only the amounts that they are willing to trade, but also the equilibrium exchange rate: 1 pound of beef is equal in utility to 10 pounds of corn. Jevons summarized it thus: Each trader must "derive exactly equal utility from the final increments, otherwise it will be for his interest to exchange either more or less." An essential notion in this quote is "final increments." It is irrelevant that at the outset, the corn-trader would have been willing to give up 20 pounds of corn for the first pound of beef. The deciding factor is the utilities of the very last pounds of beef and corn that the traders are willing to exchange. Actually, the intention is even more rigorous. Strictly speaking, *final* refers to the very last fraction of the last ounce of the last pound that is being traded. Hence, when speaking of *marginal utility*, the allusion is to the utility of the infinitely small amounts of commodities at the margin of the decision whether to exchange goods. The reference to the infinitely small is the reason that differential calculus—which, following Newton, Jevons called "fluxional calculus"—comes into play.

There is much more to *The Theory of Political Economy*, of course; there are chapters devoted to labor, rent, and capital, and many more instances where differential calculus is applied.

Jevons's life was tragically cut short on August 13, 1882, just a few weeks before his forty-seventh birthday. Jevons had gone to Hastings, in the south of England, for a summer vacation with his family. That Sunday morning, while walking along the beach with his wife and children, he mentioned that he would like to go for a swim. His wife tried to dissuade him because locals considered the seashore dangerous for bathers, but Jevons persisted. He left the family on the beach and went to the house where they were staying, apparently to change. Shortly after eleven o'clock, four boys came running toward a passer-by, crying out that someone was drowning out at sea. The man ran toward the spot indicated by the boys and saw a body floating in the water, about forty yards from the shore. Jevons had been a good swimmer, but the tide was on the ebb, and the icy coastal waters had carried him out to sea, where he drowned.

The family had been no stranger to personal tragedy. Jevons' mother had died when she was only fifty. After she passed away, the eldest son, Roscoe, went insane at sixteen and died at the age of forty. Shortly thereafter, his sister Henrietta lost her mind as well, and brother Herbert, in New Zealand, died at age forty-two. And recall that his father had died when Jevons was far away, in Australia.

Unbeknownst to Jevons (at least at first), someone else over on the continent was working on the very problems that he was tackling. Léon Walras (figure 4.2), a Frenchman born in 1834, was a year older than Jevons. His father, Auguste Walras, born in 1801, had been a promising student at *École Normale* in Paris, the most prestigious institute of higher learning in France. Like most alumni of the *École Normale*, Auguste, who had taught himself the fundamentals of economics, as they were known then, nurtured academic aspirations. For a short while, he did serve as a lecturer in philosophy at the University of Évreux, a provincial town in northern France, but the necessary step up, to one of the major French universities in the center of the country, remained closed to him. One reason was that, upon getting married to the daughter of a local family, Auguste had promised his parents-in-law that he would not take his wife away from Évreux. Another was that academic positions in economics were usually awarded not to men of science, but to businessmen and politicians who had a certain standing in society or who had family connections to the powers that be.

Lacking both, Auguste had to forgo an illustrious university career. Instead, he took on the post of headmaster of the local school and confined his scientific activities to writing articles about economics while watching his former classmates, like Antoine Cournot (more about him later), amass academic honors. Only late in life, after failing to gain a State Doctorate (the French authorization to teach at university), was he awarded a professorship, and that was based on his publications alone. The post was at the University of Caen, again in the north of France, far from the centers of learning. But even there, the freethinking man of science soon clashed with the ignorant clerics and

FIGURE 4.2: Léon Walras.

Source: Wikimedia Commons

ministers who ran the universities. Nevertheless, he fulfilled his duties diligently, both as lecturer and university administrator, and when an official letter arrived at his house one day, he hoped that the state was about to award him the Légion d'Honneur for his services. He could not have been more wrong. Upon breaking the red seal on the envelope, he learned its true content. It was a letter of dismissal. He was discharged in disgrace.

Auguste continued to write, publish, and give lectures until his death in 1866, but he had little to look back on except a missed career. (After his death, his doctor would say that he had died of a broken heart.) In 1908, Léon seized the opportunity to settle old scores with his father's enemies and gave a loving tribute to his father, by publishing his biography. In the end, Auguste's greatest contribution to the science of economics was that he passed his lifelong passion on to his son, Léon.

As a young man, Léon was unsure whether he should pursue a scientific or a literary career. He had studied both mathematics and literature in high school. But when he tried to gain entry to *École Polytechnique*, the best engineering school in France, he failed the entrance exam . . . twice. The efforts did, however, bear some fruit. In preparation for the second attempt, Léon studied a book by his father's classmate Antoine Cournot—probably at the father's behest—entitled *Recherches sur les principes mathématiques de la théorie des richesses* (Research on the mathematical principles on the theory of wealth), which had been published in 1838. It was his first encounter with mathematical economics. Rejected by Polytechnique, Léon entered the *École des Mines*, a technical institute for future mining engineers.

The prospects did not appeal to him, though, and for a while, his literary ambitions prevailed. Bitterly disappointed by the failed Revolution of 1848— King Louis-Philippe had been replaced by a president, Louis Bonaparte, who shortly thereafter installed an authoritarian regime, with himself as Emperor Napoleon III—Léon was going to pursue his ideals by writing sociocritical novels. In 1858, at age twenty-four, he published his first attempt, *Francis Sauveur*. But then his father stepped in. As befits the son of a family "*petit bourgeois, très honorable, très catholique, très royaliste*,"[7] Léon worshipped his father. There was not even a hint of Oedipal revolt in him; he would abide by whatever his father wished. But Auguste, the experienced schoolmaster and lecturer, was not about to simply impose his will on the young man. Instead, he appealed to his reason. One day, on a walk through the countryside, he pointed out to his son that idealistic authors concerned about society were a dime a dozen. What the world sorely lacked, however, were social scientists. Léon saw the light and, on the spot, promised to give up literature and devote himself to the study of economics.

But Walras had to make a living, so he started out as a journalist, writing on economic subjects first for the *Journal des économistes*, and later for the tabloid *La Presse*. The income, though meager, did allow him to set up house with a woman by the name of Célestine-Aline Ferbach, a single mother whom he later married and whose son he adopted. He had a daughter with her, whom the couple named Marie-Aline.

Unwilling to moderate his socialist inclinations, he lasted less than a year at *La Presse*. For example, in a two-part exposé, published in October 1860, he ascribed the high rents for apartments in Paris, and the high cost of living in general, to the large proportion of luxury residences in the French capital. This perceived evil provided him the occasion for a bitter attack against the "sterility of exaggerated luxury," which leads, he proclaimed, to the production of fireworks and jewelry instead of food and clothing.[8] Critical articles like these did not endear him to the conservative readers of the paper, and his reformist ideas also displeased the owner. Walras was soon fired. He subsequently found employment as a clerk at the *Chemins de fer du Nord*, the railway company in the north of France. The job provided him a steady income, but the busy schedule left him no time for scientific work.

After three years, he resigned from the railroad to take a managerial position at a financial institution with the catchy name *Caisse d'escompte des associations populaires de consommation, de production et de crédit*, (literally: Discount bank of the national associations for consumption, production and credit), one of several banks that were founded to advance funds to small businesses. It seemed that attending to financial matters would provide a better fit with Walras's inclinations toward economics, and he, his mother, and his sister even invested in this financial institute. Alas, the bank became insolvent, he lost his job, and the family members lost their money. As Walras explained later, one reason that the *Caisse* went bankrupt was that its inexperienced agents, sitting face to face with prospective borrowers, were prone to grant credit where it was not warranted. "A real banker," he wrote, "would have said to himself, 'this venture could fail, I won't do it.'"[9]

Walras found himself unemployed and in debt. Fortunately, he had come to the attention of Joseph Hollander at the *Caisse*, one of its external overseers. Hollander offered him a position at his private bank, Trivulzi, Hollander, and Cie. He was now thirty-six years old.

A year and a half later, he got his real break. He had given a talk in 1860 at an international conference on taxation in Lausanne, the Swiss town on Lake Geneva. His remarks about taxation as an instrument of social justice had made quite an impression, and the Swiss Council of Public Education recommended him for a teaching position at the Academy of Lausanne (now the University of Lausanne). Nothing came of it then, but ten years later,

Louis Ruchonnet, then a member of the Council of the Canton of Vaud and later president of the Swiss Confederation, remembered him.

Cantonal officials had decided to establish a chair of political economy in the Academy's law department and, urged on by Ruchonnet, they encouraged Walras to apply. It was not an easy decision for him. He would have to give up his position at Hollander's bank, and if the appointment in Lausanne did not come about, he would be out of a job entirely. But Walras did not hesitate; here was a chance to realize his life's ambition. He sent his publications, along with a proposal to teach mathematical economics, to the jury that had been appointed: three public figures, who favored his appointment, and four professors, three of whom were dead set against it. But the fourth professor, Henri Dameth, an open-minded economist from the University of Geneva, cast the deciding vote. Like his colleagues, he did not share Walras's views. But in spite of his reservations, he held the firm opinion that in the interest of science, Walras's ideas should be taught. And so, with four votes against three, he was appointed. However, because of the professors' concerns about his socialist leanings, his initial nomination was only as an extra-ordinary professor, and only for one year. Fortunately, before that year was over, the position was converted to a regular, tenured professorship.

But there was a further stumbling block. When the position was offered to Walras in December 1870, the Franco-Prussian War was in full swing. All physically eligible Frenchmen between the ages of twenty and forty had to register with the authorities and confirm that they were available to take up arms. To obtain his travel documents, Walras presented himself at police headquarters accompanied by two city councilors, former classmates of his, who asserted that he would return to France if he were called up. Now he was good to go.

His inaugural lecture in Lausanne reflected the prevailing uncertainty. "This day ought to have been the happiest of my life," he declared wistfully, before expressing the fear that he might not be able to teach for long, before being called to the colors. "Thus, perhaps in a few days I may have to leave this chair which I today occupy for the first time. If such be the case, remember me with sympathy and kindness."[10] But fate was kind to Walras. The war passed him by, and he occupied Lausanne's chair of political economy for twenty-two years.

Walras's most important work, *Éléments d'Économie Politique Pure, ou Théorie de la Richesse Sociale* (Elements of pure economics, or the theory of social wealth), was first published in two volumes in 1874 and 1877. Further editions followed in 1889, 1896, and 1900, and several more appeared posthumously. The book, which is considered a classic in the history of economic thought, contains a wealth of ideas. However, because our interest here is rather specific, I will mention only in passing what many consider to be Walras's greatest achievement—namely, his theory of general equilibrium.

This theory asserts that through commerce and trading, all markets for commodities will eventually clear, in the sense that prices adjust to levels where demand just equals supply. Walras gave a mathematical proof of this claim by positing a system of equations that express demand and supply, with prices being the unknown variables. By counting the equations and the variables, he showed that the system is, in principle, solvable. This means that prices exist that solve the equations, and hence the markets clear. The question remained, however, how these prices are reached. Walras maintained that one commodity would appear, the so-called *numéraire*, which would serve as the common denominator for all markets. All prices would be expressed in terms of this *numéraire*, which is, of course, what we know as *money*. Then, through a dynamic process that Walras called *tâtonnement* (groping), traders approach the correct prices through trial and error. In the course of several rounds of adjustment and fine-tuning, the prices iterate toward values that clear the markets.

A defining factor of Walras's work is the mathematical treatment that he bestows upon his subject matter. His father, Auguste, influenced by Augustin Cournot, had touted just such an approach to economics as far back as the early 1830s. The latter's *Recherches sur les principes mathématiques de la théorie des richesses* was arguably the first published attempt to introduce mathematics into the study of political economy. By *mathematics*, Cournot did not just mean the use of symbols and arithmetic operations for accounting purposes. He referred to the scientific rigor and strict logic of mathematics, which permit the discovery of hidden connections among variables. In his words: "Those versed in mathematical analysis know that [mathematics] is not only used for the purpose of calculating numerical magnitudes; that it is also used for finding relationships between variables... [and] functions."[11]

Walras had had the idea of creating a mathematical theory of economics for many years. In the letter in which he originally offered his services to the Academy of Lausanne, he had expressed his intention to put political economy on a mathematical footing. Now was the time to make good on that promise. Unfortunately, he was somewhat hampered in this attempt by his father's at-best semirigorous definitions, to which he adhered (or at least that he did not want to disavow, possibly out of filial respect). For example, Walras *père* had used the vague term *rareté* (sparsity or scarcity), to denote what Walras *fils* revealed to be none other than marginal utility. It is "the intensity of the last need to be satisfied by a quantity of a good that is possessed," he wrote, in a slightly convoluted fashion, in the first edition of the *Éléments*.[12] What he meant was that if someone owns, say, 30 pounds of a good, *rareté* refers to the utility that the 30th pound bestows on the owner. In *Principe d'une théorie mathématique de l'échange* (Principle of a mathematical theory of exchange), published already in 1874 in the *Journal des économistes*, Walras had been even more specific. "*La rareté*," he asserted, "is the derivative of effective utility with respect to the quantity possessed, exactly as one defines speed as the derivative of the covered distance with respect to the time that was needed to cover it."[13] Voilà—the mathematical definition of marginal utility.

Apparently, the fact that a commodity's marginal utility decreases as its quantity increases seemed so obvious to Walras that in the first edition of his *Éléments*, he mentioned it only in passing. He began his explanation by arguing that beyond a certain quantity, consumers would not want any more of a good, even if its price were zero, because at that point, all their wants have been satisfied. In other words, as quantity increases, the utility for the good drops towards zero. And after pointing out that the utility of the good is the underlying reason that the demand curve slopes downward as its prices increases, he proclaims that "one has to accept, I repeat, that rareté increases when the quantity possessed decreases."[14] Thus, he formulated the familiar principle, albeit in the opposite direction of the one to which we are accustomed. He did add "and vice versa," which makes his statement more familiar: Marginal utility decreases as quantity increases. Fifteen years later, in the second edition, he was more specific when discussing the accumulation of a commodity: "All subsequent units... from the first unit which fills his most urgent want to the last after which satiety sets in, have a diminishing intensive utility for him."[15]

Walras made no mention of Daniel Bernoulli in his *Éléments*, but he must have been aware of his essay, written 150 years earlier. His theory of exchange is based on the principles that people strive to maximize utility and that utility diminishes as the quantity that is possessed increases. Let us say that a merchant wants to exchange meat for barley. He will do so, as long as his marginal utility of meat is lower than his marginal utility of barley multiplied by the price of barley (where the price is expressed in units of the goods).

Let me illustrate. Suppose that the market price for 1 pound of meat is 8 pounds of barley. At the outset, the merchant has lots of meat, but no barley. Hence, his appetite for barley is great (high marginal utility), while his demand for meat is satiated (low marginal utility). So he will buy barley, and his total utility will rise somewhat. In fact, as long as his marginal utility for meat is lower than eight times his marginal utility for barley, the merchant will buy more and more barley, and his total utility will continue to rise. Will the process ever stop? Yes, because implicit in Walras's theory is the principle of decreasing marginal utility: In the course of trading, the merchant's stock of meat diminishes, and thus its marginal utility rises, while his stock of barley rises and its marginal utility falls. Hence, there must come a time when the marginal utility of meat has risen and the marginal utility of barley has fallen, to reach the point where the ratio is exactly 8:1. That is when the merchant's utility is at a maximum, and he stops trading.

Similar reasoning applies to the use of raw materials in the production of goods. Even if, say, the same amount of chocolate is required to produce the first and the last chocolate chip cookie, the utility of chocolate to the producer does not remain constant. The more cookies he produces, the lower the cookies' price; and as a result, the marginal utility of chocolate decreases. Hence, the producer will exchange chocolate for, say, flour, so long as his total utility increases. He will stop trading when the marginal utilities reach the point where they are proportional to the exchange rate.

Walras realized that that there was much more work to be done on his magnum opus before it could be published. In the meantime, he wanted to inform the interested public about his discoveries, and—at least as important to him—to establish priority. He sent reprints of his paper on mathematical economics from the *Journal des économistes* to various economists abroad. One of the designed recipients was Stanley Jevons in England. Walras had heard of the professor in Manchester and believed him to be a mathematician who applied mathematical tools to statistics. Surely he would also be interested in a mathematical treatment of economics. So, on May 1, 1873, he sent him a copy of his paper.

Just a few days later, he suffered an unpleasant surprise. An astute twenty-four-year old law student at Leiden University in the Netherlands, Johan d'Aulnis de Bourouill, had just read that very paper in the journal. The young man was greatly impressed, but the manner in which Walras approached economic questions with mathematical tools seemed familiar. He recalled having read something similar elsewhere, and on May 4, he wrote to the professor in Lausanne to tell him about a book by an Englishman named Stanley Jevons, published two years earlier: *The Theory of Political Economy*.

It was a horrible blow to Walras. Here he was, laboring in solitude on what he believed would be a groundbreaking theory, while someone else over in England had already worked out everything. Any claim to priority was shattered. After barely a week, still smarting from the bad news, he received a thank-you letter from Jevons, posted from Manchester on May 12. It confirmed his worst fears; everything he had heard from d'Aulnis de Bourouill was true. Jevons's tone was laudatory, albeit with just a whiff of haughtiness. Heaping faint praise on Walras's work, Jevons mainly extolled his own. "You will find, I think, that your theory substantially coincides with and confirms mine," he declared and added, condescendingly, "the publication of your paper as it now stands is very satisfactory in so far as it tends to confirm my belief in the correctness of the theory." Just in case Walras might claim that he had worked out his theory several years earlier, Jevons made sure to assert that his ideas went back even further: "All the chief points of my mathematical theory were clear to my own mind by the year 1862, when I drew up a brief account of it, which was read at the meeting of the British Association of Cambridge."[16]

Jevons was right, of course. Within a time span of just a few years, both he and Walras had come up with identical theories—Jevons first, his colleague in Lausanne second. "What you call the rarety of a commodity appears to be exactly what I called the coefficient of utility at first, and afterwards the degree of utility, which, as I also explained, was really the differential coefficient of the utility considered as the function of the quantity of commodity." Toward the end, Jevons got down to the crux of the matter. Rubbing salt into Walras's wound, he wrote that the latter's paper "might lead to misapprehension as to the originality and priority of its publication. I shall therefore take it as a favour if you will kindly inform me whether you are sufficiently acquainted with my writings or whether you would desire me to forward to you a copy of my 'Theory of Political Economy.'"[17]

Walras was devastated. Chagrined, he responded on May 23. He meekly noted the remarkable parallels between his and Jevons's works, and after an all-but-desperate attempt at finding at least some differences—if not in substance then in the approaches—he launched into an epic account of how his work had come about. He stressed that the only sources of inspiration for his findings were his father and Cournot. He quoted his father's pronouncement of 1831, "political science is a mathematical science,"[18] and his comparison between speed, time, and distance on the one hand, and *rareté*, quantity, and utility, on the other, and Cournot's book, which dated from 1838. He himself started thinking about these ideas in 1860, but he was forced to interrupt his research for pecuniary reasons, only being able to pick up again ten years later. In minute detail, he recalled when and how he pronounced his findings and triumphantly pointed to the positive reception that his ideas had had in Italy. Promising to make Jevons's book known to the public wherever an occasion to do so presented itself, Walras ended his letter with a pitiful plea, asking Jevons to bring his own ideas to the attention of his students and readers as well.

Hidden behind these polite phrases was a deeply disappointed man. His innermost feelings became apparent two months later, in another letter to Jevons: "I make no pretense of being above human frailty. I confess to you frankly that I was at first quite upset by the loss of my priority." But admirably, he kept his composure, saying, "When there is nothing else left to do but yield, one might as well do so with good grace."[19]

The only thing he could do, in fact, was to get his work into print as fast as possible. He decided to publish his book in two parts so that the first, which contained his theory of diminishing marginal utility, would not have to wait until everything else was ready to go. Volume 1 of *Éléments d'Économie Politique Pure* (Principles of political economy) was published in 1874, three years after

Jevons's *Theory of Political Economy*. It contained the generalized marginal theory across the economy (i.e., in consumption, production, and exchange). The second part, which contained further theories on production, capital and credit, and tariffs and taxes, came out three years later.

Jevons too was concerned with priority. In 1879, in the preface to the second edition of *The Theory of Political Economy* he politely, but just a wee bit dismissively, asserted that the work of Walras was important "not only because they complete and prove that which was before published elsewhere...but because they contain a third or fourth independent discovery of the principles of the theory.... The fact that some four or more independent writers...should in such different ways have reached substantially the same views of the fundamental ideas of economic science, cannot but lend great probability, not to say approximate certainty, to those views." In speaking of "independent writers," Jevons referred not only to himself, but also to a few other obscure scientists who had done groundbreaking work, as we shall see in the next chapter. Later in the preface, he added that "the working out of a complete system based on these lines must be a matter of time and labour, and I know not when, if ever, I shall be able to attempt it."[20] This remark stung Walras especially sharply. He felt that Jevons was willfully (and unfairly) ignoring his work—after all, he already had worked out a complete system.

Unfortunately, Walras's income at the University of Lausanne never sufficed, and monetary problems were a constant worry to him. When he gave up his position with Trivulzi, Hollander, and Cie in Paris, he took a salary cut from 4,000 francs per year to 3,600 francs, which was the customary wage for professors in Lausanne. His former wage had not allowed him to accumulate any savings, and before he could take up his new post, he had to ask the university for an advance on his salary to pay for his trip. Even then, he did not have enough money to pay for his family's transportation to Lausanne; they had to wait in France until a later date. As a professor, he incurred great expenses publishing and distributing his works, which were exacerbated by his wife's serious illness—she died in 1879 at the age of forty-five. To augment his income, Walras taught extra classes, contributed articles to the *Gazette de Lausanne*, and served as consultant to the insurance company *La Suisse*.

In 1884, Walras remarried. His second wife was Léonide Désirée Mailly, a fifty-year-old woman of some financial means. With this and a nonnegligible inheritance from his own mother, his financial worries diminished. But the strain had taken a mental toll, and in 1892, at age fifty-eight, Walras took early retirement. The decision may have been made easier by the knowledge that the brilliant Italian Vilfredo Pareto would succeed him in the chair.

His retirement did not mark the end of Walras's scientific efforts, however. If anything, he threw himself into his work with increased vigor. *Études d'économie sociale* appeared in 1896, and *Études d'économie politique appliquée* in 1898. In 1900, Walras's second wife died. After that, together with his daughter Marie-Aline, he moved to Clarens, a small town near Montreux, where he continued his work. In 1905, he had a brain wave. Just ten years earlier, Alfred Nobel had established the Nobel Foundation. The first Nobel Prizes were awarded in 1901, and the initial recipients of the Peace Prize were the founders of the Red Cross and various pacifists, several of them Swiss. Walras decided that he deserved the Nobel Peace Prize. Why not use economics to further world peace?

The bylaws of the Nobel committee clearly specify that candidates for the prize are not permitted to nominate themselves. So Walras coopted three colleagues from the University of Lausanne to write a letter of recommendation. To ensure that his views on how to achieve world peace were explained correctly, he himself wrote the memorandum that accompanied the letter. The proposition that he advocated was that the abolition of taxes and tariffs would encourage free trade across borders, and free trade across borders would lead to world peace. Governments and states would finance their activities through the ownership and rental of real estate.

Alas, it was not to be. First, the nomination, mailed on July 20, 1905, arrived after the deadline for nominations. His candidature had to be postponed until the following year. Then the advisor charged by the committee to evaluate the submission gave only a lukewarm endorsement. He did stress the importance of Walras's work in mathematical economics, but he paid scant attention to his proposal to further world peace. It may have been the advisor's way of saying that the work, though of great interest for science, was not worthy of the Nobel Peace Prize. So Walras's nomination was passed over, as were twenty-seven others, and the 1906 Nobel Prize for Peace went to President Teddy Roosevelt instead.

A final high point of Walras's career came in June 1909, when the University of Lausanne organized a jubilee celebration to honor his fifty years of "unselfish work" to further science. Walras spent the last years of his life ordering and revising his and his father's works, in the conviction that one day, their collected writings would be published. He died on January 5, 1910, at age seventy-five.

Occasionally, two thinkers, working independently, happen upon a deep result simultaneously. The discovery of infinitesimal calculus by Isaac Newton in England and Gottfried Wilhelm Leibniz in Germany is one example; Charles Darwin's and Alfred Russell Wallace's development of the theory of evolution is another. In economics, the options pricing formula was developed in 1973 by a team consisting of Fisher Black and Myron Scholes, on the one hand, and, by Robert Merton on the other.[21] It is rarer, however, that three scholars, working in different countries, writing in different languages and talking to different audiences, should independently and more or less simultaneously discover something fundamental. Such was the case, however, when Stanley Jevons in Manchester, Léon Walras in Lausanne, and now our third hero, Carl Menger (figure 4.3) in Vienna, discovered the power of mathematical economics and realized that it is utility (particularly marginal utility) that underlies economic decisions.

Menger was five and six years younger, respectively, than Walras and Jevons. If Jevons's manner was professorially dull, and Walras's appearance was that of a self-effacing, polite scholar, Menger had an abrasive style and self-assured demeanor that added excitement to the normally staid milieu of German-speaking scholars.

His life started ordinarily enough. Born in 1840 into minor nobility, Carl Menger grew up as the middle of three sons on an estate in Galicia in the Austrian Empire (now Poland). His father was a lawyer, his mother the daughter of a wealthy, bohemian merchant. On the family estate, he still witnessed serfdom before it was abolished in 1848 (one of the few achievements of the revolution whose outcome had so disappointed Walras). All three brothers, whose full family name was Menger Edler von Wolfensgrün, studied law at the University of Vienna. Carl got his doctorate of law at the University of Cracow before returning to Vienna to obtain his *habilitation*,

FIGURE 4.3: Carl Menger.

Source: Vienna University Archive

the authorization to teach at a university, in 1872. Both of his brothers became famous: Anton, a year younger than Carl, a professor of law concerned with social justice, developed into an ardent socialist, and Max, two years older, became a member of Parliament, where he was known as the voice of small businesses and tradespeople.

In 1863, while still a student preparing his doctorate, Carl started working as a journalist at the *Amtliche Lemberger Zeitung*, the administrative paper in

Lemberg (Lviv, today in Ukraine). He continued his journalistic career at the *Wiener Zeitung* in Vienna, the official gazette of the Austrian government. As a staff member of this periodical, Menger was now a member of the press department of the Austrian prime minister's office, a civil servant. One of his responsibilities was to analyze markets and write reports about them. The experience had a profound influence on his understanding of the economy. He noticed that the experts with whom he talked, as well as the merchants themselves and their informants, ascribed the movement of commodity prices to factors other than what the classical textbooks advocated. Thus, he began to research, think, and write. His efforts culminated in 1871 in the publication of *Grundsätze der Volkswirtschaftslehre* (Principles of economics). In 1873, at the relatively young age of thirty-three, he was appointed to a chair in economic theory at the University of Vienna. He kept his position as a civil servant because the tuition fees that the thirty or forty students paid to attend his classes hardly sufficed as a salary. Two years later, in 1875, he left the *Wiener Zeitung* to take up a new and uniquely challenging position: private tutor to Rudolf, crown prince of Austria-Hungary.

Because the scions of European royalty did not attend public schools, they had to be instructed privately. Franz-Josef, the emperor of Austria and king of Hungary, attempted to have his son educated in the military tradition by the stern—even sadistic—General Leopold Graf Gondrecourt. The six-year-old boy, who had been promoted to the rank of colonel upon his birth, was routinely woken in the morning by pistol shots, was forced to take cold showers and perform hours of military drills in pouring rain, and was abandoned at night in a forest. Rudolf was a sensitive and scared little boy, and the horrifying drills had a detrimental effect on his physical and mental state. To make things worse, Rudolf's mother, Empress Elizabeth, known to her adoring subjects simply as Sisi, suffered from psychosomatic disorders that she kept under control by traveling and keeping a cool distance between herself and the court, and her son.

But when a whistleblower informed her of the harsh regime that the boy had to suffer through, she demanded in no uncertain terms that the general be fired, and that Rudolf be given an education that accorded with his inclinations. Even the powerful monarch could not defy an ultimatum by his wife, and the heartless general was dismissed. A much more liberal-minded military man, General Josef Latour von Thurmburg—the actual whistleblower—was henceforth tasked with Rudolf's schooling. He hired dozens of tutors, some of high standing in academia: the superintendent of the Imperial Natural History Museum, Christian Gottlieb Ferdinand Ritter von Hochstetter, the president of the German Ornithological Society, Eugen Ferdinand von Homeyer, the eminent zoologist Alfred Brehm, and, of course, the professor of political economy at the University of Vienna, Carl Menger.

What started as a student-teacher relationship developed, little by little, into a deeper friendship. Menger mentored the crown prince, nurturing his liberal tendencies against his father's autocratic style of governing. In the lessons, which lasted between nine and fourteen hours a week, the professor impressed upon the prince the principle that the people's interest in their own welfare is the basis of a sound economy. Therefore, they must be given the freedom to choose and act according to their own will; a state cannot replace activities driven by private interests.

After a year of study, the nineteen-year old prince and the professor set out on an educational journey through Europe, North Africa, and the Middle East to gain an understanding of economic relationships. A book that Rudolf wrote about his voyage through the Middle East immediately earned him a PhD at the University of Vienna, a degree which, he admitted himself, he would never have received for such meager proof of scholarship as his travelogue. So now he was not only a colonel, but also a *Doktor*. Another work, this one on ornithology, earned him honorary membership in the Imperial Academy of Sciences. After two years of such scholarship, the prince's guardians had had enough of Menger's (and the other tutors') liberal influence on the bookish Rudolf, and Emperor Franz Josef put an end to it. The tutors were let go, and from then on, Rudolf was occupied with hunting deer, riding horses, breeding dogs, and chasing women.

The unfortunate young man was torn between his liberal inclinations and the conservative politics at his father's court. His love life was equally torrid. Apparently, the only woman he had ever really loved was a fetching Polish Jewish woman, with whom he was not allowed to associate. Forced to marry Princess Stephanie of Belgium, whom even Rudolf's mother referred to as a *hässliches Trampeltier* (ugly camel), he became a womanizer and indulged in various imperial amusements. The end came soon. One evening in January 1889, on a retreat to his hunting lodge in Mayerling, in Lower Austria, he retired to his bedroom with his seventeen-year old mistress, Baroness Mary Vetsera.

When his valet came to wake him the next morning, there was no answer. Together with Rudolf's hunting companion, they forced open the door and found the lifeless bodies of the baroness and the prince. Rudolf had shot his mistress in the temple and then killed himself.

The news took a while to reach his parents in the Austrian capital. A personal envoy was dispatched by special train to Vienna, where, according to protocol, a certain lady-in-waiting had to be summoned because only she could inform the empress of the terrible news, who, in turn, was the only person allowed to break the news to the emperor. Interestingly, the stock exchange in Vienna knew about Rudolf's death hours before the emperor and his wife—a station chief at the railroad stop in Mayerling had telegraphed the news to Nathaniel Meyer Rothschild, head of the eponymous bank.

Rudolf's death changed the course of history. Because he had been Emperor Franz-Joseph's only son, it was Franz Joseph's brother, Karl Ludwig, who was next in line. But Karl Ludwig renounced his rights in favor of his son, Franz Ferdinand—who was famously assassinated in 1914, thus starting World War I. Because Franz Ferdinand had married a countess who was not of a family sanctioned by the Habsburgs for matrimony with one of their own, his children were removed from the line of succession, and it was Franz Ferdinand's nephew Karl who eventually became Emperor Charles in 1916, the last emperor of Austria and last king of Hungary, the very last royal of the Habsburg dynasty.

When he was still in charge of Rudolf's education in economic and social matters, Menger, the former journalist, taught Rudolf how to write opinion pieces for newspapers. In the following years, the crown prince had published dozens of anonymous essays about the court's intrigues in the *Neue Wiener Tagblatt*, mostly critical of his father's politics. The paper's editor, Moritz Szeps, became a close confidant, though, given the rampant anti-Semitism in Austria at the time, the crown prince's friendship with the Jewish editor had to be kept strictly secret.

The most remarkable piece, apparently produced jointly by Rudolf and his professor, was a pamphlet entitled *Der österreichische Adel und sein constitutioneller Beruf: Mahnruf an die aristokratische Jugend* (The Austrian nobility and its constitutional mission: Appeal to the aristocratic youth). Published anonymously in 1878 by "an Austrian," it castigated the sons of the nobility as being lazy and incompetent, unable to complete a school curriculum, and unwilling to do any kind of useful work. Disinclined to participate in any kind of public service or parliamentary activity, the young aristocrats occupied themselves in the spring with hunting, horseback riding, and visiting neighbors; the summer was whiled away amid great boredom in Switzerland, in the Austrian mountains, or at spas; fall was devoted to the planning of balls, soirées, and theater performances; and the winter was spent attending them.

As an example of the pervasive and utter incompetence, the anonymous publication ascribed the Austrian army's recent poor showing in battle against the Prussians to the fact that Austrian noblemen came to their officers' ranks (if they bothered to obtain them at all) by excelling in chivalry and horsemanship rather than by studying military doctrine and the art of war. Rudolf naturally knew what he was talking about; after all, he had obtained the rank of colonel as a baby. One may assume that the piece was not composed entirely by the twenty-year old crown prince, but that Menger had also had a hand in its writing.

Let us now return to Carl Menger. After obtaining his law degree in 1867, while writing for various publications, like *Presse, Botschafter, Debatte, Allgemeine Volkszeitung, Wiener Tagblatt* (which he founded), *Wiener Zeitung,* and *Neue Wiener Zeitung,* Menger was also hard at work on his *Grundsätze der Volkswirthschaftslehre,* which appeared in the same year as Jevons's *Theory of Political Economy,* and three years before Walras's *Éléments d'Économie Pure.*

Impressed by the way that experts traded on the markets, Menger lamented the inability of economic science, as it was then taught, to analyze and explore the foundations of economic behavior. The fact that practitioners routinely threw aside hitherto-developed economic principles in order to rely solely on their instincts and experience was not due to flippancy or incompetence. Their professed lack of interest in the then-current economic theories, he writes, must stem from the sterility of all attempts to gain any useful empirical knowledge from them.

In the first chapter, Menger discusses the characteristics that an object must possess to be recognized as a good. Four prerequisites must be fulfilled: A human need exists; the object is able to satisfy this need; people recognize that this object does, in fact, satisfy the need; and, finally, the object is at one's disposal. Prerequisites two and three are tantamount to saying that the object must provide utility to its owner. There are aberrations, of course—objects that seem to be goods even though they are not, like amulets or snake oil, which fulfill only imaginary needs or are thought to satisfy a need but do not. On the other hand, there are goods that are intangible, like copyright, patents, goodwill, and monopoly power.

The relationship between goods and the economy as a whole are discussed in the second chapter. In order to prosper, people must be able to foresee their needs and act accordingly. While the "wild Indians" only care about the next few days, Menger wrote, nomads are one step up because they foresee and provide for their needs several months ahead. "Civilised people," on the other hand, the *crème de la crème* of humanity, predict the goods that they require not just for the coming winter but further into the future; they make provisions for the years to come and even plan for the lifetime of their offspring. He discusses how people determine their future needs, how they compute the quantities they require in each time period, and what activities they undertake to satisfy these needs in the most efficient manner.

It is in the third chapter that Menger turns to the subject of our interest here (namely, the utility of a good). A set of needs might be satisfied by a certain good, he writes. To determine the good's value, these needs must be listed in order of diminishing importance. Then one utilizes the amount of goods that one possesses to fulfill as many needs as possible, starting with the most important and continuing down the list. Menger illustrated the matter with a farmer who requires grain for all kinds of necessities: He uses the first batch to make bread in order to keep himself and his family above the subsistence level. The next batch will be used to make other baked goods in order to ensure continued good health. He will use the batch after that as seeds for next year's crop. If some grain is still left over, the farmer will make beer and brandy with it. Any remaining grain will be used to feed his pets.

Now, if the farmer has insufficient grain to satisfy all these needs, he will not reduce each need in proportion—eat less, bake less, and so on— but, rather, just stop feeding the pets. Hence, the last need that *can* be satisfied—the need to make beer, in this example—determines the grain's value. If farmers have barely enough grain to keep their families from starvation, grain is very valuable. If they have enough of it to feed their dogs, it has not much value.

Of course, this example represents precisely the theory of diminishing marginal utility, as it was expressed simultaneously by Jevons and, later, by Walras. The more of an article or commodity one possesses, the less one would pay for an additional unit because it would convey only a little utility. A good's value, therefore, is determined by the utility of the last marginal unit. A good provides the greatest amount of utility when it is utilized to ensure survival, less when it is employed to obtain the means for well-being, and least when it is used to access luxury. Of course, because all units are identical, there is no first, second, or third unit. Hence, given the amount available, all units will be valued identically: highest if there is only enough of it to guarantee survival, lower if it exists in sufficient quantity to ensure well-being, and least when it is so abundant that it can be used to access luxury.

With the account of the farmer and his grain, Menger expressed one of the main themes of his *Volkswirthschaftslehre*, and indeed of his teaching. So important was the topic to him that he did not settle for presenting just one example. In the pages following, he restates the principle of diminishing marginal utility over and over again . . . and in great detail. He admits that some readers may be bored by his repetition, but he insists that the price is worth paying if his demonstrations gain in clarity. Accordingly, he elaborates about water on a rocky island, biscuits on a boat lost at sea, living space in apartments, gold, and diamonds. By the time he gets through discussing all these cases, the principle of diminishing marginal utility should be abundantly clear to all readers.

Interestingly, Menger employed not a single equation or mathematical symbol in his 285-page work. In contrast to Jevons and Walras, he remained firmly wedded to the tradition of framing his theory via narrative and anecdotes. When considering the last unit of a good, for example, he simply refers to the last pound of grain or the last bucket of water. He never considered infinitesimally small amounts at the margin, which would have inevitably led to the use of calculus. The lack of mathematical sophistication is unsurprising because Menger, trained as a lawyer, was obviously not very conversant in mathematics and stuck to notions that nonmathematicians like himself could grasp. By the way, this is the reason that he required so many examples to make a point that could have been driven home more easily with the help of a single graph or equation.[22]

Nevertheless, Menger's description of diminishing marginal utility is closely associated logically with the mathematical method that was evolving in the 1870s, and this is why he is considered one of the founding fathers of *marginalism*. When Jevons compiled a bibliography of "Mathematico-Economic Writings" for the second edition of his *Theory of Political Economy*, he included Menger's *Grundsätze* in the list, even though the book was mathematical only in spirit. As a later economist would put it, Menger's work is one of the writings "whose mathematical character is not explicitly expressed in symbols or diagrams."[23] Similarly to the title character of Molière's *Le Bourgeois gentilhomme*, who never knew that he had been speaking prose all his life, Menger did not realize that his principle of diminishing marginal utility was quite mathematical in nature.

In Lausanne, 800 kilometers to the west, Léon Walras learned of Menger's work only in 1883. Once again, it was the industrious and well-connected Johan d'Aulnis de Bourouill who brought the news. In a letter dated June 22, 1983, he told Walras of a professor in Vienna who was well acquainted with Walras's work. He had just received from him the very interesting book *Grundsätze der Volkswirthschaftslehre*, twelve years after publication, which he said captivated by its clarity, logic, and simplicity of exposition. Walras, always eager to disseminate his own work, availed himself of the opportunity to send one of his own publications to Menger, who returned the favor by dispatching his *Grundsätze* by return mail. In the accompanying letter, he confirmed that he had been following Walras's work for years, but then he launched into a harangue about the mathematical method as used in economics. He did not subscribe to the mathematical method, he fumed, which could only serve, if at all, for the purposes of demonstration, not for research. At most, it can help to discover quantitative relationships. Economic laws may be clothed in mathematical terms or demonstrated by graphical presentation. Thus, mathematics can be quite useful, but it does not come close to the essence of research and can be no more than an auxiliary discipline of political economy. Rubbing salt into the wound, Menger ends his letter rather condescendingly, pointing out that a large number of men have been pursuing this line of research for many years but are not cited anywhere in Walras's work.

Walras responded quickly. Thanking Menger for his letter and the book, he pointed out that, in spite of their differing methods, they both follow a "rational" approach to the study of political economy. He suggested that they join forces in order to overcome the resistance with which their approach was being met in Germany, where empiricism reigned supreme—the concept that knowledge can be acquired not through logical deduction, but solely through sensory experience. Concerning the remarks about mathematics, Walras played down their differences. After all, he wrote, the connections between prices on the one hand, and the factors that determine them on the other, were precisely the quantitative relationships that Menger was willing to allow for. Moreover, he saw no difference between mathematics as a tool for research and mathematics as a method of exposition. He himself used mathematics for explanatory purposes, after having utilized it for research. Indeed, he wrote, it was mathematics alone that allowed him to clarify the questions with which he dealt. In summary, he did not consider himself a mathematician who endeavored to employ equations at all cost, but an economist who strove to obtain a reasoned opinion that could be rigorously proven. With the help of mathematics, he was successful, he claimed.

The respectful and polite, if at times acerbic tone that Menger employed toward Walras belied a fierce and cynical streak that had been evident since *Appeal to the Aristocratic Youth*, that he coauthored with Crown Prince Rudolf. Menger seemed to relish controversy, as did an academic opponent in Berlin...

The resistance in Germany to which Walras alluded in his letter to Menger was more than just a minor academic disagreement. It was a major controversy that would bear upon the study of political economy in Germany for many years. The preeminent social scientist in German academia at the time, the reigning high priest, was Gustav von Schmoller, one of the proponents of the so-called historical school. Menger's view, which stated that universally valid economic laws can be deduced in a logical manner from first principles or axioms (e.g., the theory of diminishing marginal utility), stood in stark contrast to the then-prevailing theory as put forward by Schmoller and his disciples. Their view, sometimes called *synthetic*, maintained that economic events are brought about through the interplay of history, geography, local traditions, psychology, and other idiosyncratic factors. Economic behavior would have to be scrutinized case by case, and after observing many similar events, one may deduce with a certain degree of confidence that the similarities express a truth, and general conclusions can be drawn from them.[24]

The dispute between the proponents of the two approaches—better described as a slugfest of insults—began in earnest in 1883, after Menger published *Untersuchungen über die Methode der Socialwissenschaften* (Inquiries into the methods of the social sciences). The bitter quarrel, the likes of which have rarely been seen in academic disputes, would later become known as the *Methodenstreit* (dispute about methods). Menger's philosophy, as formulated in his book, was that one has to isolate first principles in an abstract and theoretical manner in order to gain an understanding of the underlying mechanism of economics, and then formulate general laws that derive from them. Noise factors that may influence economic behavior to some degree—transaction costs, windfall profits, price bubbles, or extraordinary events—should be disregarded, in the same manner as physicists ignore friction in their first approximations. Conclusions based on historical evidence, even if substantiated by statistical evidence, must be rejected.

The dispute was not just a hairsplitting academic quibble; its implications held great importance. On the one hand, if Schmoller's view were correct, and economic laws depended on history, then institutional and political design would be an important regulator of the economy. On the other hand, if Menger's view were correct, and economic laws derived from first (human) principles, then institutions and regulations would have no long-term effect. Self-interest would supersede all regulatory efforts. Even revolutions would make no difference.

In the preface of *Untersuchungen*, Menger, who obviously relished controversy, gave ample, if underhanded credit to the German scientists whose work he was about to attack. He was fully aware, he wrote, that his polemical style would ruffle some sensibilities, but he believed that it was indispensable to shake the proponents of so-called historicism out of their complacency. After years of unreflected high-handedness and a concomitant disregard for any critical assessment of their theories, a pointless phraseology about the methods of political economy had emerged, absent any meaning. Under these circumstances, he opined that his polemical style was warranted.

In his charge that the historical school suppressed any countervailing opinions, Menger was correct. As the acknowledged apostle of German sociology, Schmoller had the authority to assign posts, monitor appointments to professorships, and keep doubters on the sidelines. And now this nonentity—from Austria, of all places—dared challenge his dominance? Schmoller was not about to let that happen. He responded with a scathing review entitled *Zur Methodologie der Staats- und Sozial-Wissenschaften* (On the methodology of political and social sciences). Right from the start, he announced that he too would not hold back because Menger had attacked him personally in his polemical writing. Sure enough, phrases like "scholastic exercise," "abstract schemes," and "mental tuberculosis" pervaded the review.

Schmoller did not negate the need for general theories, but he maintained that descriptive accounts like travelogues, reports about trade shows, and statistical data were necessary prerequisites. To leave aside such important factors would be to misjudge the most elementary principles of scientific methodology. True, he conceded, conditions are never replicated precisely, and hence no general theory can be deduced. But the belief that some basic principles—like human needs and self-interest—could lead with mathematical precision to general laws about social science was naïve, he claimed, and best left to otherworldly armchair scholars. To ignore all other motives was tantamount to ignoring reality and making do with fictitious decision-makers. Admittedly, abstraction is required in science, but one must abstract correctly in order to arrive at scientific truths, and not chase after dreamy robinsonades and unreal phantoms, as Menger allegedly did. It was pretentious of him, said Schmoller, to start with indefensible hypotheses and to believe that the most complicated phenomena would be explicable in terms of a single factor. Menger was not familiar with, or was consciously disregarding, progress that has been made in psychology, linguistics, philosophy of law, and ethics, which led to the comprehension of many phenomena of the masses. Instead, he adhered to the mystic belief that economics could be understood as a game of egotistical interests. On the other hand, he may simply lack the "organ" to understand the essence of the historical school, he concluded. Truly, Schmoller was not one to mince words.

Toward the end of his review, after a dismissive "we are done with the book," he deigned to dispense lukewarm praise. Menger was a keen dialectician, Schmoller conceded, a logical thinker, an extraordinary scholar. But he lacked comprehensive philosophical and historical training and did not possess the breadth of mind that would be required to absorb experiences and ideas from different viewpoints. "He is correct in much of what he reproaches the historical school of German economists . . . but he is certainly no reformer."[25] Being intimately familiar with only a tiny corner of the science of economics, Menger believed it to be all there was.

Up to that point, Schmoller had kept the tone of his review acerbic but low-key, as befitted a scholarly piece. But now he let his dislike of Menger get the better of him. He was not annoyed by Menger's views, he blustered, which, after all, simply enlivened scholarly debate, but he did resent the schoolmasterly tone with which Menger scolded everybody who dared have an opinion that differed from his. Schmoller's resentment of the Austrian upstart was palpable.

Menger was not about to take such insults lying down. Launching into an 18,000-word rebuttal, he paid Schmoller back, and with interest. The title of the pamphlet already set the tone. Framed as a collection of sixteen letters to an imaginary friend, *Die Irrthümer des Historismus in der deutschen Nationalökonomie* (The errors of historicism in German political economy) permitted Menger to hold forth about Schmoller's philosophy. There was not much new in the rejoinder except that the tone of voice was raised by several octaves. The peculiar literary style that he chose allowed Menger to refute objections that Schmoller raised, could have raised, or might have raised. The venerable German professor was mentioned by name no less than 139 times.

And how was he mentioned? Menger spared no words. Schmoller's arguments were unreflected, uninformed, plucked out of thin air, full of invectives, irresponsible, incomprehensible, barbaric, and replete with misunderstandings, misrepresentations, and errors. He subscribed to confused notions and befuddled concepts; his language was vulgar, unseemly, and half repulsive, half absurd. He confused even the simplest notions of science, was indiscriminate in the choice of means, and lacked understanding. His knowledge being limited to a primordial soup of historical-statistical material, he abused scientific publications and acted like a one-sided party hack, transferring the bad habits of political fights to scientific discussion.

As polite academic protocol required, Menger sent Schmoller a copy of his booklet, inscribed with the words "From the author." Did he truly hope that Schmoller, editor of the influential *Jahrbuch für Gesetzgebung, Verwaltung und*

Volkswirthschaft im deutschen Reiche (Annals of legislation, administration and political economy in the German Empire), would review it in his journal? Well, Schmoller did not. In fact, he did worse than that. He returned the book unread, along with a very rude letter, in which he said that he would not lower himself to such a task. And as if that were not enough, he topped it off with what must have been a first in German scholarly debate: He published the refusal verbatim in the *Jahrbuch*! The exquisitely rude tone of the "review," rare even in a very heated academic discourse, merits its full, translated, reproduction here:

> The editorial office of the *Jahrbuch* is unable to report on this book, since it has already been returned to the author with the following remarks: Dear Sir! I have received by mail your booklet "The Errors of Historicism in German Political Economy." It carries the handwritten note "from the author," which means that I have to thank you personally for forwarding it. Various sources had informed me for some time already that this pamphlet contains mainly attacks against my person, and a first glance onto the first page confirmed this. As much as I appreciate your preoccupation with me and your willingness to enlighten me, I do want to stay true to my principles concerning such verbal exchanges. I will now reveal them to you and can only recommend them for imitation; one is thereby able to spare oneself much time and anger: such personal attacks are tossed, sight unseen, into the fireplace or the wastebasket, especially when I cannot expect any furtherance for myself from this author. In this way, I will never be tempted to bore the public with the prolongation of such feuds, as some German professors do, who engage in stage combats. However, I do not want to be so impolite, as to destroy a book arranged so nicely by your hand. Therefore, I return it to you with thanks and with the request to make better use of it elswhere. By the way, I will always be very grateful for further attacks, because, as the saying goes, "many enemies, much glory."²⁶

The most significant innovation that Jevons, Walras, and Menger introduced into the discourse on economics was that decision-makers strive to maximize their total utility and, as Daniel Bernoulli already had proposed 150 years earlier, the utility provided by a good to its owner diminishes as the amount already owned increases. The term that is nowadays used for this phenomenon is *diminishing marginal utility*. It was coined by Friedrich von Wieser, one of Menger's most brilliant students, who called it *Grenznutzen*, German for "utility at the frontier", or "at the margin."

Menger had repeatedly referred to political economy, as practiced by Schmoller and his disciples, as "German.'" It was meant in a disparaging manner, to indicate that his colleagues in the north were intellectually stuck in ancient, obsolete paradigms that led nowhere. In response, Schmoller coined the expression *Austrian School*. It was meant as a dismissive term, designating provincialism and backwater science. In an admirable exercise in branding, Menger's disciples reversed the meaning, and *Austrian* was soon taken as a sign of distinction. Nowadays, many carry the label *Austrian School* as a badge of honor.

Menger's conviction that historical data give no indication as to how the economy works, and that it is therefore futile to design economic institutions and set up regulations, appeals to laissez-faire enthusiasts. The idea that it is men and women themselves who shape their economy, solely by maximizing their own utilities without interference by any government, appeals to libertarians, beginning with the writer Alisa Zinoyevna Rosenbaum, better known as Ayn Rand, by way of economists Ludwig von Mises and Nobel Prize winner Friedrich Hayek, all the way to congressmen and former presidential and vice-presidential candidates (respectively) Ron Paul and Paul Ryan.

CHAPTER 5

FORGOTTEN PRECURSORS

While Stanley Jevons and Léon Walras were busy discussing priority—Carl Menger preoccuppied by the *Methodenstreit*, seems to have stayed aloof from that issue—two new names popped up all of a sudden: Jules Dupuit in Paris and Hermann Heinrich Gossen in Cologne, and the question of who had preceded whom became moot.[1] Dupuit, a Frenchman born in 1804, had studied mathematics at the *École Polytechnique* and engineering at the *École Nationale des Ponts et Chaussées* before becoming a chief civil engineer in a French *département*, and later chief engineer of the Paris water works. He made a name for himself not only by designing roads and bridges—among other tasks, he was involved in work on the Champs-Elysées—but also for supervising the construction of the sewage system in Paris.

After witnessing a major flood on the Loire River in 1846, Dupuit turned his attention to flood management. His claim to fame in the realm of economics was an article he published in 1844 in the *Annales des Ponts et Chaussees*. Entitled *De la mesure de l'utilité des travaux publics* (On measuring the usefulness of public works), he introduced to engineering problems what is today known as *cost-benefit analysis*. Along the way, he illustrated how investment funds should be allocated through a financial technique that is now called *capital budgeting*. He resorted to an example of water usage, exactly as Menger would do thirty years later. But unlike Menger, Dupuit started with a downward-sloping demand curve and deduced from this the principle of diminishing marginal utility. As he was the first to state the concept in the realm of economics, he was entitled to do it in any way he wanted.

Dupuit began by declaring that demand curves slope downward: "As the price of an item decreases, more and more consumers will make use of it, in addition to the previous consumers who now consume it in greater quantity." He saw no need to justify this stipulation. In a somewhat surprising statement coming from an engineer trained in mathematics, he claimed that "this is an experimental fact so often confirmed by statistical observation that we have no further need to prove it."[2] From this, he deduced the principle of diminishing marginal utility, illustrating it thus: At a flat rate of 50 francs per year for daily delivery of a hectoliter of water, a man who owns a home orders 1 hectoliter; at a price of 30 francs, he orders 2 hectoliters. Hence, the first hectoliter has a utility of at least 50, because he agreed to buy it at that price. The additional hectoliter has a utility of less than 50 (because he did not buy it when the price was that high), but the utility is at least 30 (otherwise, he would not have agreed to buy more water after the price dropped). So, Dupuit continues, at a price of 50 or 30 francs, the homeowner will buy only enough water for his personal needs. But when the price falls to 20 francs, he will wash his house every day; and at 10 francs, he will also water the garden; at 5, he will dig and fill a pool; and at 1 franc per hectoliter, he will buy so much water as to ensure a continuous flow. Voilà—the principle of declining marginal utility.

Published in a French engineering journal, Dupuit's work was not well known outside France. It did lead to a response by a colleague in the *Annales des Ponts et Chaussées* in 1847 and a long follow-up by Dupuit two years later. Apart from that, it did not elicit much interest. Only Walras (who, after all, spoke French) was aware of Dupuit's groundbreaking work. But after having discovered, to his chagrin, that Jevons's work had preceded his, he was loath to admit that his ideas had been anticipated by yet another thinker. In his attempt to downplay his predecessor's work, however, he made a mistake. Eager to disperse any possible suspicion that he had been influenced by the French engineer, he himself inadvertently alerted Jevons to the latter's existence in a note in 1877. Jevons, honest scientist that he was, lost no time—he immediately made Dupuit's work known to other economists with whom he corresponded frequently. In the preface to the second edition of his *Theory of Political Economy*, he gives him full credit: "It is the French engineer Dupuit who must probably be credited with the earliest perfect comprehension of the theory of utility."[3]

The revelation of Dupuit's discovery was but a prelude to what would follow. While the principle of diminishing marginal utility was implicit in Dupuit's work, albeit disguised as an engineering problem, it was explicitly stated in the context of political economy in a much ampler work that was completely ignored after its publication. The author, Hermann Heinrich Gossen, son of a Prussian tax collector, was born in 1810; he became a civil servant at the urging of his father. His real interest was political economy though, and, starting in his mid-twenties, he set out to investigate the rules that govern human interactions. In 1854, he published a book entitled *Entwickelung der Gesetze des menschlichen Verkehrs, und der daraus fliessenden Regeln für menschliches Handeln*, which may be translated as "Development of the laws of human commerce, and of the resulting rules of human action." This was ten years after Dupuit's paper, but still twenty years before Jevons, Walras, and Menger published their supposedly groundbreaking works.

Gossen was no believer in humility. "What Copernicus accomplished to explain the interaction between the heavenly bodies, that, I believe, I achieve in explaining the interaction between people on the surface on the earth," he wrote in the preface. "My discoveries have enabled me to indicate with absolute certainty the orbits people must follow in order to achieve their mission in life in the most fulfilling manner."[4] Truly, no false modesty here.

He ended the preface by expressing his wish that the public would give his book a critical but unbiased reading. The latter request was especially pertinent because his work demolished many erroneous ideas that until then had been considered to be true. He himself had been forced to abandon them only with great pain, he wrote. They had become dear to the public, and even more so because many men had staked their professional reputations on their truth. If they now abandoned these ideas, they would find themselves in the same position as he had found himself—namely, to be forced in the prime of life to seek a new position. Truth be told, he never liked being a functionary of the state anyway, and after his father's death he abandoned his career in the civil service. Thenceforth, he made a living through the sale of insurance policies to farmers against the threats of hail and mortality of cattle.

Gossen held the firm belief that mathematics is essential to the study of political economy. Indeed, he was convinced that the imbroglio that pervaded the discipline was due to lack of mathematical treatment. Two essential maxims guided his inquiries. The first was that over their lifetimes, people strive to maximize the sum total of their utility, which he called "pleasures." The second was that the additional utility provided by a particular good diminishes the more one consumes of it, until satiation is reached.[5]

Of course, this maxim, nothing less than the principle of diminishing marginal utility, had been known since Daniel Bernoulli's time. But with Gossen being the first author to employ this principle specifically in the realm of political economy, it is now often called Gossen's First Law. He illustrated it with three rather strange examples before moving on to more familiar ones. The enjoyment of a piece of art diminishes, he wrote, the longer one contemplates it; thinking about a problem becomes increasingly boring the longer one remains preoccupied with it; and explaining a discovery is at first exhilarating, but then it turns into tedious teaching, and finally just dull work. Nearly as an afterthought, he turned to the more customary example of consumption of bread or beef.

Building upon the First Law, he spun his thoughts further, thereby becoming the first person to formulate a fundamental economic principle that is now often called Gossen's Second Law, and is colloquially expressed as "getting the most bang for the buck": when a person decides which of several goods to purchase, he will pick the one that provides most utility for his money. Because the utility of additional amounts diminishes, the next time he makes a pick, a different good may give him more "bang," and he will choose that one. He will continue to make his picks in this manner until he has spent all available money. At that point, the additional utility that he would obtain if he had one more dollar to spend would be identical, no matter on which of the available goods he would decide to spend it.

Gossen was full of awe before the wisdom of the Lord, who had created not only the laws of nature, which held the universe together, but also the law of diminishing utility, which regulated human interaction. Even though He had granted humanity free will, He had arranged it, in His infinite wisdom, such that the one obstacle that would seem to stand in the way of attaining the common good (namely, the egotism of humans) does not thwart, but rather maximizes the community's well-being.

It was a groundbreaking discovery, and as his reference to Copernicus in the preface testifies, Gossen was fully aware of this. Alas, the book was problematic. Generously endowed with graphs and equations, it lacked some crucial elements. It had no chapters, no captions, and not a single heading. The entire book was one long argument that had to be read from start to finish—all 278 pages of it—at one time. And even the mathematics, which should have clarified things, were considered at the time an impediment. Mind you, the math was simple—nothing fancy like calculus or differential equations, except for the occasional determination of a maximum or a minimum—it was just arithmetic, pages and pages and pages of it. To make things worse, he presented numerical examples in interminable tables in an effort to aid those readers who dislike even basic mathematics.

The long-winded text was no simpler. Sentences comprising seventy or eighty words abounded, enough to put off even the best-disposed reader. Unable to find a publisher, Gossen had the book printed in 1854 at his own expense. But it found few buyers, and fewer readers. In 1858, he removed it from circulation, destroying the remaining copies. Deeply disappointed, he died the following year of tuberculosis, at age forty-seven.

It took another twenty years for Gossen's forgotten work to gain the recognition it truly deserved. While Jevons and Walras were having their debate on priority—with Walras ceding to Jevons—the thought actually arose that maybe, unbeknownst to them, others had also hit upon the same idea. It was a colleague of Jevons in Manchester, one Professor Robert Adamson, who found the first indication that such was in fact the case. A brief reference to Gossen's work in a book was all he could go on. He advertised a search for it, but for a long time was unable to obtain one of the rare copies. Only when a mention of it appeared in a German bookseller's catalog in 1878 could he finally order a copy.

It did not take long for Jevons, who was unfamiliar with the German language and had to rely on Adamson's description and oral translations, to realize, once again, that his priority over Walras had been but a fleeting victory.

"Gossen has completely anticipated me as regards the general principles and methods of the theory of Economics," he admitted in the preface to the second edition of his *Theory of Political Economy*.[6] In an attempt at damage control, he pointed out that Gossen assumed, for the sake of simplicity, that a person's utility diminishes in a linear fashion, while functions in economic science are rarely completely linear. After listing several other purported deficiencies, he went to some lengths to dispel any possible suspicion that he may have been inspired by Gossen. "I desire to state distinctly . . . that I never saw nor so much as heard any hint of the existence of Gossen's book before August 1878."[7] To bolster his point, he cited a number of personalities who also had never heard of Gossen—even though, as he indicated in a footnote, a copy of his book had been acquired by the library of the British Museum in 1865. Wistfully, though, he had to conclude that "from the year 1862, when my theory was first published in brief outline, I have often pleased myself with the thought that it was at once a novel and an important theory. From what I have now stated in this preface it is evident that novelty can no longer be attributed to the leading features of the theory. Much is clearly due to Dupuit, and of the rest a great share must be assigned to Gossen."[8]

Walras, for his part, took the matter in stride. After having learned that Jevons, Menger, and Dupuit had preceded him, it was no great shock to him to discover yet another precursor. In fact, he took it upon himself to right what he perceived as a scientific wrong. May be he also felt just a hint of gleeful satisfaction that Jevons, whose detection had caused him such distress, had at least been anticipated himself by an unexpected pioneer.

Walras made inquiries and came upon the only apparent relative of Gossen's, a nephew by the name of Hermann Kortum, who was a professor of mathematics at the University of Bonn. Kortum provided detailed information about the Prussian state assessor's life, and Walras published a twenty-five-page tribute to the forgotten thinker in the *Journals des économistes* in 1885. In his letter to Walras, Kortum expressed little astonishment at the silence with which his uncle's book had been received. "This lack of success was no surprise in a country where in spite of a row of excellent mathematicians, from Euler to Riemann and Weierstrass, mathematical culture never took hold among professionals apart from astronomers, physicists and a small number of engineers, and where, even today, the sight of an equation will cause the majority of your colleagues to flee."[9]

How did Menger take the news about Gossen's work? He apparently was never even aware of it. Caught up in the here and now of his *Methodenstreit* with Schmoller, he remained oblivious to all else that may have taken place in the past. The great Schmoller himself, however, put off by Gossen's insistence on utility as the basis for decision-making, dismissed him without further ado as an "ingenious Idiot" whose book had led to all kinds of mischief. Coming from the leading proponent of the historical school, someone so despised by Menger, the unwarranted and unjustified criticism may just count all the more as praise.[10]

CHAPTER 6

BETTING ON ONE'S BELIEF

One of the first people to take up the theory of diminishing marginal utility in a setting that truly deserves the qualification "mathematical" was the Cambridge philosopher and mathematician Frank Ramsey (figure 6.1). The contributions of this extraordinary man during his all-too-short life—he died before the age of twenty-seven—would have sufficed to fill the careers of several scholars, not only in his chosen subjects but also in the field of economics. Ramsey was born in 1903, the eldest son of a minor academic, an author of textbooks on mathematics and physics who later became president of Magdalene College at Cambridge. The mother, a graduate of Oxford, was an ardent feminist and a pillar in Frank's life. The family had two more daughters and a son. The latter, Michael Ramsey, would become Archbishop of Canterbury—the only sibling who remained a Christian. One sister would become a medical doctor, and the other a lecturer in economics.

Frank showed his mettle from an early age. Entering Trinity College at Cambridge with a full scholarship, he graduated at age twenty as Senior Wrangler, the university's top-ranked student of mathematics. But he had made a name for himself even before then, with an exceptional feat. Largely self-taught in the German language, the eighteen-year old undergraduate translated Ludwig Wittgenstein's notoriously difficult *Tractatus logico-philosophicus*, one of the most important philosophical treatises of the early twentieth century, from German into English. (Wittgenstein, the youngest son of a fabulously wealthy Austrian family, who had come to Cambridge to study under the philosopher

FIGURE 6.1: Frank Plumpton Ramsey.

Source: Art by Patrick L. Gallegos (2017), courtesy of Wikimedia Commons

Bertrand Russell in 1911, wrote the first version of *Tractatus* in the trenches of the battlefields of World War I, during which he distinguished himself as a courageous officer, and later in Italian captivity. In the fall of 1923, Ramsey traveled to Austria for a fortnight to discuss the *Tractatus* with Wittgenstein himself, who was then teaching schoolchildren in the village of Puchberg.)

A year later, Ramsey fell in love with Margaret Pyke, a feminist and birth control activist. Unfortunately—for Ramsey at least—Margaret was married. She was the wife of a journalist and inventor who today would be described as a "nerd" or "gearhead" because of his unconventional and often unimplementable inventions. But this did not dishearten Ramsey. Intellectually mature, way beyond his years, he remained boyish at heart. Once, after a walk with her along a lake, when they were resting (she reading a book, he pretending to do so), Ramsey posed the decidedly nonphilosophical question, "Margaret, will you f*** with me?" Apparently, she was in no mood to consent, replying, "Do you think once would make any difference?"[1] The indifference of her response, even though quite appropriate, must have struck him as unwontedly coldhearted.

His unreciprocated love for a married woman ten years his senior drove him to the newest fad among the intellectuals of the early twentieth century—psychoanalysis. He traveled to Austria again, this time to undergo analysis in Vienna with one of Sigmund Freud's first disciples, the psychologist Theodor Reik. One can only surmise that his desire for an unattainable, significantly older woman must have spoken volumes about Frank's relationship to his mother, whom he adored and who would die four years later in a car accident.

Cured of his doomed love, Ramsey returned to Cambridge. Thanks to the efforts of the eminent economist John Maynard Keynes, he was elected a Fellow of King's College in 1924 and was made a lecturer of mathematics two years later. At the same time, he got involved with a student of psychology, Lettice Baker, who would later become a distinguished portrait photographer. To abide by the mores of the times, or at least to hide their transgressions, they were forced to keep their frolicking secret. Afraid of being found out by either his mother or by the Fellows at King's College, they had to sneak off to Lettice's room at Trinity College for their trysts.

Ramsey's fears were well founded. Keynes had informed him that even his previous liaison with Margaret could constitute an impediment to his election as Fellow.[2] Eventually, Ramsey's mother found out about him and Lettice, and this hastened the young couple's decision to get married. But theirs was not to be a traditional union. They agreed that they would be free to pursue other conquests, and they both took advantage of their liberty. Frank understood this to mean that he had to report to Lettice his innermost feelings . . . toward other women. Two years after the wedding, he fell in love with one of Lettice's friends, Elizabeth Denby, a social reformer who, like Margaret, was nearly ten years older than he was.

Ramsey was rather better at dishing it out than at taking it. Once, while he was spending a Christmas holiday in France with Elizabeth, Lettice took up with an Irish writer and, as was their custom, informed Frank of the liaison. He was furious. There he was, doing nothing more than spending a little quality time with his mistress, all the time looking forward to resuming a tranquil life with his wife, and there she was, destroying it all. Considering himself the wounded party, expecting compassion, he lamented plaintively, "I had decided to give up Elizabeth, and you know how important she is to me." Somewhat belatedly, he arrived at the conclusion that "I can't stand the strain of this sort of polygamy and I want to go back to monogamy." His wish was soon fulfilled, in fact, because the affair between Lettice and the Irishman came to an end after a few weeks. When she informed him how miserable she felt, he did not bother to hide his glee and relief. She responded with sarcasm: "Well, it's something that my being unhappy has helped you . . . Let me cheer you by saying I'm still very gloomy and depressed."³

Wittgenstein was lured to Cambridge by Keynes and Ramsey in 1929. In spite of his fame, the Austrian philosopher still had no academic degree, so he needed to submit a thesis before being allowed to teach. Ramsey, now a twenty-six-year-old with only a bachelor of arts degree, became the nominal PhD thesis supervisor of his forty-year-old friend. This was after their relationship had undergone a severe test: During one of their conversations, Wittgenstein had expressed the opinion that Freud, though very clever, was morally deficient. After that, the two refused to speak to each other for quite a while. The thesis that Wittgenstein eventually submitted was none other than the *Tractatus*, a first version of which had previously been considered as insufficient to award him a bachelor's degree because it had no preface and no footnotes. One of the examiners at the successful thesis defense was Bertrand Russell.

The following year, Ramsey fell severely ill. He had developed jaundice that was initially thought to have been caused by a blockage of the gallbladder. An unsuccessful operation at a hospital in London revealed that he had long suffered from a liver and kidney disease without being aware of it. He died a few days after the operation, leaving behind Lettice and two daughters. One can only speculate how philosophy and mathematics would have advanced, quite apart from economics, had he lived longer.

John Maynard Keynes, Ramsey's mentor at King's College, was to become the most influential economist of the twentieth century. He had distinguished himself as a civil servant and as the representative of Britain at the Versailles peace conference after World War I. The twenty-three-year old Ramsey came to his attention in 1926 with the paper *Truth and Probability*, in which he took issue with Keynes's own *Treatise on Probability*, published five years earlier. By extending the familiar deductive logic of definite conclusions to an inductive logic of partial conclusions, Ramsey defines probability theory not in the way that physicists or statisticians do, as a frequency or as the proportion of favorable outcomes to all outcomes, but in the spirit of the seventeenth-century German polymath Gottfried Wilhelm Leibniz, as a branch of logic—namely, "the logic of partial belief and inconclusive argument."[4]

Following that paper, and with Keynes's encouragement, Ramsey set his formidable mind to economics. For a philosopher of his talents, it was no more than a simple diversion. He wrote and published only two papers in this field: *A Contribution to the Theory of Taxation* in March 1927, and *A Mathematical Theory of Saving* in December 1928. Keynes, who had them published in his *Economic Journal*, was fulsome in his approval. In an obituary for Ramsey, he wrote, "The latter of these [referring to *A Mathematical Theory of Saving*] is, I think, one of the most remarkable contributions to mathematical economics ever made."[4] Although the paper was certainly of the highest caliber, this praise is somewhat exaggerated. If anything, the paper was one of the very first in the field of economics that really deserved the moniker *mathematical*. Hence, "the most remarkable contribution ever made" sounds suspiciously like a tautology. In his general assessment, however, Keynes was right, even if it took half a century, until the 1970s, for the paper's groundbreaking importance to sink in.

Nonetheless, it is the paper *Truth and Probability* that is of interest to us here. Unfortunately, Ramsey's writing style was somewhat convoluted at times, and a bit pompous. For example, when referring to Keynes's interpretation of probability, he expressed his disagreement in the following manner: "I hope that what I have already said is enough to show that [Keynes' theory] is not so completely satisfactory as to render futile any attempt to treat the subject from a rather different point of view."[5]

If probability is the degree of belief (i.e., the degree of confidence one has in a certain outcome), how can it be measured? "It is not enough to measure probability," Ramsey wrote. "In order to apportion correctly our belief to the probability, we must be able to measure our belief."[6] Because belief and confidence stem from a person's subjective impressions, a psychological approach must be taken, and the degree of belief must be measured psychologically. But how? One can agree on a few benchmark figures: Full belief in an outcome is denoted by 1.0, full belief in the opposite by zero, and equal beliefs in the proposition and its contradiction by ½. But what is meant by a belief of 2/3? That the belief that an outcome will occur is twice as strong as the belief that it will not?

One can certainly not make do with an ordinal scale, which simply indicates which belief is greater. Such a scale, like the one devised to express the hardness of materials in the early nineteenth century by the German mineralogist Friedrich Mohs, suffices to indicate which material can scratch another. As the name says, an *ordinal* scale provides an ordering—in this case, listing materials in terms of increasing hardness. Mohs simply assigned a higher number to a harder material, arbitrarily giving a value of 1.0 to the soft mineral talc and a value of 10.0 to diamonds, the hardest material then known.

But his ordinal scale had an important shortcoming. While it allowed the positioning of materials on the scale of increasing hardness, one cannot say that the hardness of quartz (7) plus the hardness of calcite (3) equals the hardness of diamonds, or that gypsum (3) is three times as hard as talc. What Ramsey sought was a way to measure belief in a *cardinal* manner, like the measurement of length or weight. On a cardinal scale, a greater number not only indicates longer length or heavier weight, but also specifies how much longer and heavier the object is. The difference between ordinal scales and cardinal scales is that the latter can be added and subtracted.

And this is what Ramsey was after. He sought a measurement process that would allow addition and subtraction. The first alternative he considered as a proxy for his ultimate goal, the measurement of probability, was to assess the intensity of feeling. The more intensely one feels that a certain outcome will occur, the higher its probability. But even before he got to checking for addability, Ramsey dismissed the option. One cannot trust a person to assign numbers to the intensities of her or his feelings. After all, even "the beliefs which we hold most strongly are often accompanied by practically no feeling at all; no one feels strongly about things he takes for granted."[7]

Ramsey concluded that in order to measure the strength of one's belief, one should measure the extent to which one is prepared to act on it. "Our judgment about the strength of our belief is really how we should act in hypothetical circumstances." To assess that strength, Ramsey took a page out of Daniel Bernoulli's playbook. "The old-established way of measuring a person's belief is to propose a bet and see what are the lowest odds which he will accept." There are also shortcomings with that approach, of which Ramsey mentions two. First, he notes that the marginal utility of money diminishes. (He mentions this only very casually, since by then this phenomenon was "universally agreed.")[8] Hence, measurements could be distorted if the potential gains or losses are so large that they would put the gambler into a different class of wealth. Also, the amount that one is willing to wager may lead to the erroneous conclusion that a person's belief in a certain outcome is stronger when he is rich than when he is poor.

The second shortcoming he cites is that people may have a special eagerness or reluctance to place bets. Surprisingly, Ramsey is a bit too nonchalant here. His statement merits—indeed requires—more than just an offhand aside because, as we already saw in chapter 1, diminishing marginal utility of money, cited by Ramsey just a few lines before, entails risk aversion. Hence, in theory, a reluctance to bet is tautologous to diminishing marginal utility, while an eagerness to bet should be ruled out as a contradiction. Ramsey does propose an explanation in terms of human psychology (namely, the joy or the dislike of excitement that accompanies gambling). In chapter 8, we will occupy ourselves at length with the dichotomy of risk aversion and the simultaneous eagerness to bet.

To avoid such weaknesses, Ramsey came up with an alternative proposal. He illustrated it with an elaborate example: "I am at a cross-roads and do not know the way; but I rather think one of the two ways is right. I propose therefore to go that way but keep my eyes open for someone to ask; if now I see someone half a mile away over the fields, whether I turn aside to ask him will depend on the relative inconvenience of going out of my way to cross the fields, or of continuing on the wrong road if it is the wrong road. But it will also depend on how confident I am that I am right; and clearly the more confident

I am of this, the less distance I should be willing to go from the road to check my opinion. I propose therefore to use the distance I would be prepared to go to ask, as a measure of the confidence of my opinion."[9] Hence, the measure of one's belief (i.e., of the probability that one is right) is the effort that one would be willing to extend in order to make sure. In Ramsey's example, the degree of belief (the probability that the chosen way is the correct one) is measured in fractions of miles.

According to Ramsey, the laws of probability are an extension of formal logic to a theory of partial belief. As such, a person's degrees of belief, proxy to the measurement of probability, must—at least in theory—be logically consistent. After all, if a measuring system is inconsistent, it is useless. Obviously, this should hold true as well for the measurement of the strengths of beliefs. For example, if my belief that "It will be sunny tomorrow" is stronger than "It will be cloudy tomorrow," and my belief that "It will be cloudy tomorrow" is stronger than "It will be rainy tomorrow," then my belief that "It will be sunny tomorrow" must be stronger than "It will be rainy tomorrow."[10] Ramsey specifies several definitions and axioms that must hold in order for the measurements to be consistent, and then he derives from them four laws of probable belief that he considers fundamental.

The first law is that the belief in an occurrence and in its opposite must add up to 1 (i.e., to 100 percent).[11] For example, my belief that "It will rain tomorrow" plus my belief that "It will not rain tomorrow" exhausts all the possibilities, and hence the expression adds up to 1. The second law can be illustrated as follows: My belief that "The Knicks will win tomorrow's game if it rains" plus my belief that "The Knicks will lose if it rains" must also equal 100 percent. The third law is a bit more involved: My belief that "I will live to be 120 years and die a billionaire" equals my belief that "I will live to 120" multiplied by my belief that "I will die a billionaire, given that I have lived to be 120." The fourth, and final, law can be illustrated as follows: My belief that "I will live until 120 and die a billionaire" plus my belief that "I will live to an age of 120 years but without a billion dollars" is equal to my belief that "I will live until 120" . . . with or without a billion dollars.[12]

Anybody who does not conform to the laws because his beliefs are not coherent is vulnerable to a spoof by sneaky bookmakers. For example, if the fourth law is violated, a shrewd bookmaker could take advantage of a naïve gambler. To illustrate, let us say that the gambler is willing to bet on the winner of a tournament among several soccer clubs. If the Grasshopper Club wins, he will receive fifty dollars. He can also place a bet on the Butterflies Club and receive fifty dollars if that team wins. Assume that the gambler's beliefs are such that he is willing to pay five dollars for each of the two simple bets. There is also a third bet, the compound bet, which pays fifty dollars if either the Grasshoppers or the Butterflies win. The gambler evaluates his belief for the compound bet at twelve dollars. Now the bookmaker could *sell* him the compound bet for twelve dollars, and *buy* the two simple bets for five dollars apiece (a total of ten dollars). She immediately pockets the two dollars that remain. If the Grasshoppers win, the gambler pays fifty dollars to the bookie because he lost one of the simple bets, but gets fifty dollars from the bookie because he won the compound bet. If the Butterflies win, he will pay fifty dollars because he lost the other simple bet, but also gets fifty dollars because he won the compound bet. If any other team wins, no money will exchange hands. In all cases, the gains and losses (if any) balance out. But the savvy bookie gets to keep her two-dollar profit, while the naïve gambler, having violated the fourth law, is sure to lose.[13]

The laws are so fundamental, says Ramsey, that "if anyone's mental condition violated these laws, his choice would depend on the precise form in which the options were offered him, which would be absurd."[14] Well . . . it may be absurd, but unfortunately it is also very common. Indeed, the manner in which a question is framed, or a choice presented, often determines how people answer, decide, and act. The laws that should, in principle, govern degrees of belief are violated much more frequently than Ramsey thought. Such violations will be discussed in part III of this book.

Ramsey's paper led to the understanding that human beings not only exhibit a subjective view of money, as epitomized by its diminishing marginal utility, but also assess probability in a subjective manner, as expressed by a person's intensity of belief. And this person's intensity of belief is expressed, according to Ramsey, by that person's willingness to act based on that belief.

However, confusion about the nature of probability continued to reign, with an abundance of interpretations. It took several more years for the commotion to settle down. In the early 1930s, the Russian mathematician Andrei Kolmogorov developed a theory of probability that was not subjective, like Ramsey's, but mathematically rigorous and objective. In the *Foundations of the Theory of Probability*, published in 1933, Kolmogorov stipulated three axioms for the measure of probability: nonnegativity (probabilities are always equal to or greater than zero), normalization (something is bound to happen), and finite additivity (the probability that one of several nonoverlapping events occurs is found by summing the individual probabilities).[15]

Kolmogorov's theory was dry, bare of emotion. But humans do not conform to that ideal—if it even is an ideal. Frank Ramsey recognized that psychology must come to bear on the study of probability. But even he stipulated that humans, with all their frailty, must be rational and consistent in their personal beliefs. The fact that they are not will be the subject matter of part III of this book.

CHAPTER 7

GAMES ECONOMISTS PLAY

On December 7, 1926, while the twenty-three-year old Frank Ramsey was lecturing at King's College in Cambridge, a twenty-two-year old Hungarian postdoc delivered a talk to the mathematical association of Göttingen in Germany. The department of mathematics at the University of Göttingen, led by the renowned David Hilbert, was at that time the most famous math department not only in Germany, but in the whole world. Aspiring mathematicians from all over flocked to Göttingen in droves, much as they do to the Institute for Advanced Study at Princeton nowadays.

One of them was a young man from Hungary, John von Neumann (figure 7.1). Already considered an exceptionally gifted mathematician, he would become a towering genius, the world's leading mathematician until his early death in February 1957, at the age of forty-three. His accomplishments are so numerous and important that one cannot even start to give them their proper due. Among them are innumerable advances in pure mathematics (functional analysis and axiomatic set theory), in physics (the mathematical theory of quantum mechanics), in computer science (the first concept of a digital computer), and his involvement in the Manhattan Project.[1] And, of course, von Neumann was the father of game theory, which was the topic of the talk he gave to the association in 1926.

The paper that von Neumann delivered in Göttingen—just six weeks after graduating from ETH Zurich with a degree in chemical engineering—was published two years later in *Mathematische Annalen*, a prestigious journal

FIGURE 7.1: John von Neumann.

Source: Wikimedia Commons, Los Alamos Scientific Laboratory

of mathematics, under the innocuous title *Zur Theorie der Gesellschaftsspiele* (On the theory of parlor games). As examples of such games, von Neumann mentioned roulette, chess, bridge, baccarat, and poker. It may sound a bit absurd that a mathematician of von Neumann's abilities would deal with something so banal as parlor games. However, nothing could be further from the truth. Parlor games can shed an important light on the fundamental question of economics, von Neumann pointed out—namely, "What will a completely egotistical 'homo oeconomicus' do under certain external circumstances?"[2] Answers to questions like this one have implications not only for economics, but also for such diverse fields as political science, sociology, psychology, jurisprudence, and even biology.

On a metalevel, the rules of all parlor games are quite simple. There are three kinds of events that determine the state of a game. There can be events that occur randomly, with known probabilities, like the throw of a die, the flip of a coin, or the dealing of a card. This is the case in games like roulette or the coin-tossing game in the St. Petersburg Paradox. Their theoretical analysis belongs to the realm of probability theory, and from a decision-making perspective, they are not very interesting. In chess, on the other hand, or in the game of rock, paper, scissors, an opposing player performs moves, and one must strive to outsmart him or her. The opponent's and your own decisions influence the outcome, but the game is deterministic in the sense that nothing depends on chance. Finally, in games like Monopoly, backgammon, poker, and bridge, both chance events and opposing players' decisions influence the outcome. In each sort of game, there are payoffs if the players win, and costs if they lose.

An underlying assumption made by von Neumann was that players make payments only to each other: No money comes in, and none goes out. Hence, the amounts that change hands overall add up to zero. Appropriately, such games are called *zero-sum games*.

As we have seen by now, how the players view and evaluate their payoffs is crucial. Not so in von Neumann's paper. He specifically excluded utility from his considerations and assumed only that players are selfish. The question that von Neumann analyzed was what a player should do in a game, given that all the other players are also selfish and also strive to maximize their expected payouts.

His main result was the famous Minimax Theorem for a two player, zero-sum game. Because in this game, one player's loss is the other's gain, minimizing the opponent's maximum winnings is equivalent to maximizing one's own minimum winnings. What a rational player should do, therefore, is ensure that the minimum amount that he wins is as high as possible. Von Neumann proved that there exists a strategy that maximizes the minimal payoff of one player, thus simultaneously minimizing the opponent's maximal gain. And this is where the matter rested for about twenty years, until von Neumann teamed up at Princeton with a fellow émigré from Nazi Europe, the economist Oskar Morgenstern (figure 7.2).

Born in the German town of Görlitz in 1902—a year before von Neumann and Ramsey—Morgenstern was the son of Wilhelm Morgenstern, a bookkeeper and minor businessman, and Margarete Teichler, an illegitimate daughter of Frederick III, king of Prussia and emperor of Germany, who served in 1888, albeit for just ninety-nine days. The king left her a small fortune when he died, which Wilhelm squandered on ill-advised business ventures.

FIGURE 7.2: Oskar Morgenstern.

Source: Vienna University Archive

Morgenstern grew up in Vienna and obtained his doctorate from the University of Vienna in 1925. He considered himself "a product of the Austrian School of Economics,"[3] and his thesis was a piece on marginal productivity. Then came a three-year fellowship at Harvard University, funded by the Rockefeller Foundation. Upon his return to Vienna, Morgenstern was named professor of economics at the university. He became an active participant in the famous Vienna Circle, a discussion group chaired by the philosopher-physicist-logician Moritz Schlick. At the group's weekly gatherings Morgenstern met luminaries like Karl Menger (Carl's son), Rudolf Carnap (of thermodynamics fame), Kurt Gödel (of "incompleteness theory" fame), Karl Popper (of "falsifiability" fame), and maybe also the renowned philosopher Ludwig Wittgenstein.

At one of these meetings, Morgenstern presented the audience with an economic conundrum. He had been thinking about what influence predictions have on the predicted events themselves. And he had arrived at a surprising conclusion: General equilibrium and perfect foresight were incompatible. Morgenstern illustrated the paradox to the Vienna Circle with a page out of Arthur Conan Doyle's short story *The Final Problem*. Sherlock Holmes, pursued by Professor Moriarty, takes a train from London to Dover, but—correctly surmising that Moriarty would try to catch up with him by taking the later, but faster nonstop train to Dover—surreptitiously gets off at Canterbury, an intermediate station. Thus, he manages to evade the evil professor. But what if Moriarty had been even smarter, had foreseen Holmes's intended action, and acted accordingly by taking a train that stopped at the intermediate station? And what if Holmes had foreseen that Moriarty would foresee that he would foresee. . . . Neither of the two men could win by outthinking the other—no equilibrium could be achieved.

After a lecture on this matter, the mathematician Eduard Cech came up to Morgenstern and told him about a paper that a certain John von Neumann had published in 1928, which dealt with similar problems. It was the very paper on games that von Neumann had presented in Göttingen in 1926.

Morgenstern, the economist, was eager to meet this mathematician, but an encounter between the two had to wait another four years. As director of the Austrian *Institut für Konjunkturforschung* (Institute of Business Cycle Research), funded by the Rockefeller Foundation,[4] Morgenstern was extremely

FIGURE 7.3: John von Neumann and Oskar Morgenstern (l-r); Dorothy Morgenstern Thomas, photographer.

Source: Shelby White and Leon Levy Archives Center, Institute for Advanced Study, Princeton, New Jersey.

busy, frequently traveling to Paris, London, and the League of Nations in Geneva. In January 1938, the Carnegie Endowment for International Peace invited Morgenstern to visit four American universities. Two months into his stay, everything changed.

The infamous *Anschluss*, the takeover of power by the Nazis in Vienna, took place in March. Morgenstern was removed from his post at the university as "politically unbearable," and his institute was taken over by his former deputy, a Nazi. The dismissals were not racially motivated; rather, they were due to Morgenstern's liberal attitude, an outlook that was the only thing he had inherited from his alleged grandfather, Frederick III. Morgenstern wisely decided to remain in the United States. He received job offers from various universities but decided to accept a three-year appointment at Princeton, where he hoped finally to meet and work with von Neumann. Half his salary at Princeton was paid by—you guessed it—the Rockefeller Foundation.

Eventually, Morgenstern not only met von Neumann, but became acquainted with some of the world's foremost economists, mathematicians, and physicists, among them Niels Bohr and Albert Einstein. In particular, Bohr's preoccupation with the disturbance of an experiment by the observer, one of the fundamental problems in quantum mechanics, brought to mind how Professor Moriarty's foresight, or lack thereof, could influence the outcome of an event. And Einstein's dinner proclamations about the priority of theory over experiment also struck a chord. The relationship between Morgenstern and von Neumann required no intermediary because there was an "instantaneous meeting of minds and a spontaneous empathy."[5] Morgenstern managed to rekindle von Neumann's interest in games, a subject on which the latter had not spent much time since the publication of his original article. The two men exchanged reprints of their papers, and thus began one of the most important and fruitful scientific collaborations of the twentieth century.

Convinced of the tremendous possibilities of game theory, albeit as yet in only a rudimentary form, Morgenstern began to write a paper in which he expounded on its significance to economists. Upon a first reading of the manuscript, von Neumann remarked that it was too concise for readers who were unfamiliar with the subject. It would have to be expanded in order to make the theory accessible to nonmathematicians. Morgenstern went back to work, but when he showed von Neumann the new version, the latter was still not satisfied with it. Then he made a suggestion: "Why don't we write this paper together?" Morgenstern was overjoyed. He had come to Princeton in the hope of meeting the great mathematician, and now, of all things, they would collaborate! How much better could it get? "Here was Johnny wanting to work with me, both of us pushing into a vast new field, never doubting its challenge, difficulty, and promise."

The two scholars embarked upon their project in the fall of 1940. Speaking German between themselves but writing in English, all their work was done jointly, the handwriting on the drafts sometimes changing two or three times on the same page. But as they labored on, the paper became longer and longer. One day, von Neumann remarked, "You know, this will hardly do as a paper, not even in two parts. Perhaps we should make a small pamphlet out of it." They began making plans for a booklet of a few dozen pages, to be published in the *Annals of Mathematics Studies*. But it grew still more voluminous, and so they approached Princeton University Press with a proposal for a brochure of about 100 pages. The director was willing, they signed an agreement... and promptly forgot about the restriction on the number of pages.

Frequent meetings followed; they went for long walks, spent vacations together, talked far into the night, all the time thinking about and discussing the theory of games. Actually, it was Morgenstern who spent nearly all his time thinking about game theory; von Neumann also pondered other matters, like the atom bomb, computers, quantum mechanics, and cellular automata. They often met for breakfast and discussed what they needed to do. Then they would get together again in the afternoon and work, and in the evening, Morgenstern would type up whatever they had come up with during their meetings each day. They toiled for more than two years, until Christmas 1942, sometimes working furiously for entire weekends. Morgenstern never considered the work drudgery. On the contrary—he wrote in an account many years later that it was an unceasing pleasure to be completely immersed in the task and to be able to appreciate the joy of discovery as they went along.

Before I continue to discuss their work, I must mention an unsavory detail about Morgenstern's youth, one that most of the hagiographic accounts about him prefer to leave unsaid. As a young man, Morgenstern, who (despite his Jewish-sounding name) belonged to the Lutheran faith, was not fond of Jews at all. Several instances of anti-Semitic comments can be found in his diaries, which are now kept in the archive of Duke University. "Yesterday there was a demonstration of Jews against anti-Semitism; what an impudence," he observed in one comment.⁶ (Had he been Jewish, he would have used the word *chutzpah* instead of the German *Unverschämtheit*.) On another occasion, he referred to "this arrogant Jew-society," and on still another, he commented about someonw as being "so abhorrently judeo-liberal." When he found out that "the pig" Ludwig von Mises, a Jewish economist who owned several buildings, was his competitor for a Rockefeller fellowship, he was furious.

It may seem sweet revenge that he was made to taste some of his own medicine when Othmar Spann, an academic antagonist at the University of Vienna, spread the rumor that Morgenstern himself was Jewish, of all things. When right-wing papers also attacked him as a Jew upon his candidature for a leading position at the Association of Austrian Industrialists, he countered these so-called accusations by asking his father to ascertain that there was no Jewish blood in his family. One of his friends finally suggested, jokingly, that henceforth he should sign his work "Morgenstern, Aryan" to dispel any doubts about his heritage.

As he grew more mature, he may have revised his opinions, or perhaps he simply found it opportune to hide them, given his friendship with von Neumann and other prominent Jewish faculty members at Princeton. In any case, no anti-Semitic pronouncements are known from his later years.

At Christmastime in 1940, Morgenstern gave a lecture in New Orleans to the American Economic Association and immediately thereafter traveled to nearby Biloxi, to spend the holiday with von Neumann and his wife, Klari. Vacations were no reason to interrupt their collaboration, of course, and it was in Biloxi that they began to discuss a fundamental question: What was to be used as a payoff in games? Parlor games are usually played with worthless chips, but that just sidesteps the problem. Should the winnings be in cash? Money, which was equal for both players, also had the advantage of being freely transferable between them. Aware of the importance of the utility concept, Morgenstern was not very happy—he insisted that they dig deeper.

They knew, of course, what Daniel Bernoulli had written 200 years earlier, and were probably aware of the work of Ernst Heinrich Weber and Gustav Fechner as well. Even though they did not mention the two German scientists by name, they wrote that every measurement must be based on some sensation, such as light, heat, or muscular effort. In the case of utility, this sensation would be the perception of preference. But this would allow only an ordinal ordering (as in "I prefer apples to oranges"), not a cardinal ordering (as in "I like apples twice as much as oranges").

Cardinal orderings do not allow the addition of two utilities for the same person, nor the comparison of utilities between different people. The way that economists traditionally got around this tricky challenge was by positing so-called indifference curves: "John is indifferent between having (a) 6 apples, or (b) 4 apples and 1 orange, or (c) 3 apples and 3 oranges, or (d) 2 apples and 6 oranges. But to either one of these combinations, John prefers 8 apples and 2 oranges. Furthermore, he is indifferent between (e) 8 apples and 2 oranges, and (f) 6 apples and 3 oranges, and (g) 3 apples and 5 oranges."

Using collections of such indifference curves, as well as the prices of apples and oranges and a budget constraint, one can find the bundle of goods (apples and oranges, in this case) that John would deem optimal. But another person might have quite a different set of indifference curves, and because no monetary payoffs are involved, the real problem was again dodged. For game theory, some notion of utility was needed.

Specifically, what von Neumann and Morgenstern sought was a method to translate preferences into numbers, such that several different outcomes, each with a specific probability, could be combined into an *expected utility*. If people's preferences can be translated into numbers in such a manner with the help of a mathematical function, it would be called a *utility function*.

Let us see if that can be done. For example, if a weather forecaster announces that tomorrow there will be a 30 percent chance of 2 inches of rain, and a 70 percent chance of 3 inches of rain, the forecaster could claim that the expected rainfall is 2.7 inches, simply by computing the weighted average. On the other hand, a forecast of 30 percent chance of rain and 70 percent chance of sunshine cannot be combined into an "expected weather" forecast. Rainfall *can* be combined into an expected outcome, but weather *cannot*.

So maybe, as with the weather, no acceptable way of measuring and assigning numbers to utilities exists. Recall from chapter 1 that Daniel Bernoulli had proposed the logarithm as a utility function. But he had done so with no justification, beyond claiming that it seemed reasonable. It was a bit like pulling a rabbit out of a hat. One must decide, therefore, first of all, what is meant by *acceptable*.

Morgenstern and von Neumann recalled that the theory of heat was initially based "on an intuitively clear concept of one body feeling warmer than another, yet there was no immediate way to express significantly by how much, or how many times, or in what sense."[7] Only later did it turn out that heat had to be described by two numbers (namely, its quantity and its temperature). The former, being an energy, is additive. The latter is also numerical, with a rigid scale, but there is a huge difference. By investigating the behavior of gases and the formulation of the entropy theorem, it became clear that temperature is in no way additive. One can say, for example, that 40°F is hotter than 20°F, but one cannot say that it is twice as hot.[8]

The situation with utility is similar, they concluded. On the one hand, they would show that a scale for utility can be devised, even if the notion seems very unnumerical at first blush. On the other hand, that scale could not be a basis for numerical comparisons of a person's preferences, nor could it be used for comparisons between different persons. Like temperature, it would be something different.

In their quest to put utility on a sound footing, Morgenstern and von Neumann considered three possible events, A, B, and C, where, somewhat confusingly, C is preferred to A and A is preferred to B. This can be written, in compact form, as C > A > B. This is simple. Now compare the event A for certain, to the events C or B, each occurring with some probabilities. Let us say that event A denotes "receive an Apple," event B is "receive a Banana," and event C is "receive a Candy." Thus, Candy is preferred to Apple, and Apple is preferred to Banana. The question is then asked of a test person (let's call her Susan): "Would you prefer getting an Apple for sure to having a 50–50 chance of receiving either a Candy or a Banana?" If she says no, the probabilities are then adjusted, an 80-20 chance, a 40-60 chance, and so on, until she proclaims that she is indifferent between the alternatives. Let us say that she indicates that she is indifferent between, on the one hand, receiving an Apple for sure and, on the other hand, a 65 percent chance of getting a Candy and a 35 percent chance of getting a Banana. Then, von Neumann and Morgenstern suggested that if Susan's preference of Candy over Banana is stipulated to be 1.0, her answer to this question indicates that she prefers Apple over Banana 0.65 as much as Candy over Banana.

The reader may be excused for thinking that this time, it was von Neumann and Morgenstern who were pulling a rabbit out of a hat. But they were not simply juggling numbers. They knew that they needed to provide a justification for their numerical manipulations. They did so with a set of axioms.

Axioms are to the mathematician what basic food ingredients are to a chef. They are the essential building blocks of a mathematical theory (or a soufflé). One cannot do without any one of them, but with them, one can create a whole universe. The five Euclidean axioms are the prime example of this concept. They are everything that is needed to build all of plane geometry. On the other hand, leave one out, the parallel axiom, and a totally different universe emerges (namely, hyperbolic or elliptic geometry).

So let's see what axiom system the two professors came up with.[9] The first problem is that people cannot always make definite choices. Many readers may remember from their childhood days the infuriating interrogation by troublemaking relatives: "Whom do you love more, Mommy or Daddy?" Some kids will come up with the answer "I love them both equally." But many will be so dumbstruck that they refuse to answer—they simply cannot decide. Game theory cannot allow this to happen; it requires an unequivocal answer . . . always. So, let's make it an axiom. And thus arose von Neumann and Morgenstern's first axiom:

1. **Completeness:** For any two events A and B, one either prefers A to B, or B to A, or is indifferent between the two. (In mathematical notation: either $A > B$, or $B > A$, or $A = B$.)

Next consider the following: If Candy is preferred to Apple, and Apple is preferred to Banana, then naturally Candy is preferred to Banana. This stands to reason . . . or so you would think. Well, think again. It turns out this may not always be true.[10] Let us say that Peter, Paul, and Mary must decide what to buy for their after-dinner drink. Peter prefers Amaretto to Grappa and Grappa to Limoncello. Paul prefers Grappa to Limoncello and Limoncello to Amaretto. Finally, Mary prefers Limoncello to Amaretto and Amaretto to Grappa. Laid out in table form, we get the following:

Peter	Amaretto > Grappa > Limoncello
Paul	Grappa > Limoncello > Amaretto
Mary	Limoncello > Amaretto > Grappa

Committed as they are to democratic values, the three diners decide to go by majority rule. They take a vote, and the preferences quickly become apparent. A majority (Peter and Mary) prefers Amaretto to Grappa, and a majority (Peter and Paul) prefers Grappa to Limoncello. Based on these two rounds, they can make their decision—purchase a crate of Amaretto. But surprise, surprise: Paul and Mary protest. What happened? The most reasonable selection method was used—one person, one vote—and they still aren't happy? Do they want to change the rules midgame?

Well, they have a legitimate grumble. Paul and Mary point out that they would prefer even Limoncello, the lowest-ranked option, over Amaretto. How come? Here is the clincher: Had the three campers had a third round of voting, between Limoncello and Amaretto, a majority would have preferred Limoncello (Paul and Mary). So let them buy Limoncello and get it over with.

But wait a minute. Buy Limoncello, and Peter and Paul—yes, Paul, the guy who insisted on the third round of voting because of his dislike of Amaretto—will protest just as vigorously. They prefer Grappa to Limoncello. So here we have it: a paradox. One does not argue about taste, and Peter, Paul, and Mary have perfectly reasonable preferences. Try as you might, the final result is that Amaretto is preferred to Grappa, Grappa to Limoncello, Limoncello to Amaretto, Amaretto to Grappa, Grappa to Limoncello. . . . we could go on and on.

Such a situation is called a *Condorcet cycle*, after the French nobleman, mathematician, philosopher and politician Marie Jean Antoine Nicolas de Caritat, the Marquis de Condorcet. Whatever the choice, a majority always prefers a different alternative. In mathematical lingo, one says that "majority opinions are not transitive." What a letdown for democracy.[11] And what a problem for game theory. So again, let's revert to the proven course of action—let's make another axiom:

2. **Transitivity:** If A is preferred to B, and B to C, then A is preferred to C. (In mathematical notation: $A > B > C => A > C$.)

We return to Susan from the earlier conundrum, who indicated that she was indifferent between receiving an Apple for sure and a 65 percent chance of getting a Candy and a 35 percent chance of getting a Banana. Can Susan really indicate her indifference with such precision? How about 63 percent

Candy and 27 percent Banana? It is very difficult to make such declarations. But no worries—we simply postulate a third axiom:

3. **Continuity:** If A is preferred to B, and B to C, then there exists a probability such that the decision-maker is indifferent between B and a weighted average of A and C. (In mathematical notation: There exists a number p between 0 and 1 such that $B = pA + (1 - p)C$.)

The fourth axiom is maybe the strangest of all. It is usually called the *independence of irrelevant alternatives*. A choice between two alternatives must not be influenced by extraneous factors. If A is preferred to B, then the sudden appearance of C should not influence one's choice between A and B. The situation can be illustrated by the following scene in a restaurant.[12] "Today, we feature apple pie and brownies," the waiter informs the customer, who, after commenting on the limited choice, decides on apple pie. A few minutes later, the flustered waiter returns to inform the patron that he had forgotten to mention that the restaurant also features ice cream. "In that case, I will have a brownie," the customer announces after short reflection.

The poor waiter is completely thrown off, and rightly so. Obviously, the diner did not care one way or another about ice cream because she did not choose it even when it was offered. But its sudden availability did reverse her choice between the two other alternatives, apple pie and brownies. Our intuition tells us that this simply should not happen. One does not expect such irrational behavior in an upscale restaurant. No rational person should ever reverse a choice simply because a lower-ranked, and hence irrelevant, alternative becomes available. Right?

Wrong! There are even examples of presidential elections in which a candidate who was ranked lower than another one a priori got elected only because a third, unelectable candidate appeared on the scene.[13] To avoid such counterintuitive situations . . . guess what? Yes, let's make it an axiom:

4. **Independence:** A preference holds, independently of any other alternative. In mathematical notation: If $A > B$, then for any C, and any number α between zero and 1, $\alpha A + (1 - \alpha)C > \alpha B + (1 - \alpha)C$.

Stated in words, if A is preferred to B, then A should be preferred to B even if another event occurs with some probability. This axiom, as innocuous as it may sound, came to haunt scholars only a few years after it was devised. (See chapter 9.)

The reader should note that in order to motivate the second and third axioms, I presented situations that took recourse not in the choices of individuals, but in decisions made by collections of people through majority votes. The aggregation of individual preferences into collective choices gives rise to a host of additional problems, which I will not discuss here.[14] The examples given here simply serve to illustrate some paradoxical situations that may arise if the axioms are violated.

In hindsight, the choice of axioms seemed obvious. Morgenstern recalled von Neumann's surprise. "Johnny rose from our table when we had set down the axioms and called out in astonishment: *Ja hat denn das niemand gesehen?*'" ("Well, has nobody seen that?")[15] After establishing these axioms, von Neumann and Morgenstern decreed that anybody who violates even one of them is irrational. This may seem a wee bit harsh. Not everybody who is unable to decide whether she prefers "a cappuccino for sure" or a "26 percent chance of hot chocolate and a 74 percent chance of mint tea" is, in fact, irrational. But, being the scientists that they were, they simply stated: Individuals are rational only if they abide by all four axioms. We shall be a bit more cautious and define axiom abiders as *vNM-rational*, indicating that even people who sometimes violate some of the axioms, and would therefore be considered *vNM-irrational*, are not necessarily cuckoo in real life. In fact, in the following chapters, we will discuss many instances when people who are quite as normal as you and I deviate from vNM-rationality.

After identifying the axioms, the real work began—namely, to prove that these axioms lead to a utility function. In the first edition of their book, in 1944, von Neumann and Morgenstern did not include the proof, even though they did have one by the time they finished the manuscript. They promised that it would be published soon in a scientific journal, but that occasion never arose. It was only in the second edition of the book, three years later, that a sixteen-page appendix was added, in which the two scientists proved in a mathematically rigorous fashion that adherence to these axioms was a necessary and sufficient condition for a person to possess a utility function.

The proof is "rather lengthy and may be somewhat tiring for the mathematically untrained reader," they apologized. They were being too subtle. The proof is tiring for the mathematically *trained* reader. For the untrained reader, it is downright incomprehensible. Furthermore, even to mathematicians, the proof is "esthetically not quite satisfying" and—supreme contempt—"cannot be considered deep." Pages and pages of tedious, step-by-step derivations follow, culminating in the declaration that given a person's preferences, "there exists a mapping" (i.e., a utility function), and any two such mappings can be linearly transformed into each other.[16]

Let me expound on the latter point by comparing temperatures measured with different scales, Fahrenheit and Celsius. While 73°F is warmer than 69°F, one cannot claim that 105°F is warmer than 100°C. But because the relationship between degrees Fahrenheit and Celsius is linear—with 32°F corresponding to the freezing point of water, and 212°F to the boiling point—there is a simple way to work around the problem. To make comparisons, one simply multiplies degrees Celsius by the factor 1.8 and adds the constant 32 in order to obtain the corresponding temperature in degrees Fahrenheit. Thus, the freezing and boiling temperatures of water in degrees Celsius are 0°C and 100°C, respectively.[17]

The book's appendix accomplished what it was supposed to do—namely, to prove that all rational possess utility functions—determined up to a factor and a constant—that translate outcomes into numbers. Most important, the now well defined notion of utility functions allowed a key concept of probability theory to be passed on to the theory of decision-making. In a monetary wager, the expected outcome could be calculated by weighing all possible outcomes with their probabilities and then summing. For example, if there is an 80 percent chance of winning $50, and a 20 percent chance of winning $100, the expected outcome is $60 (= 0.80 × 50 + 0.20 × 100). If one played this wager over and over again, one would, on average, get $60 per round. In the same manner, one can weigh utilities of the outcomes by their probabilities and sum the results to obtain the expected utility. Faced with several options a vNM-rational decision-maker will choose the course of action that maximizes expected utility.

Morgenstern and von Neumann's theory establishes that everyone has a utility function and therefore can compare alternatives, and thus decide which one gives more utility.[18] However, it must be emphasized that the utilities of different individuals cannot be compared: Neither can the utilities of several people be added to provide a communal utility function. Every person possesses his or her personal utility function, which is incommensurable with anybody else's. And it is not a matter of simply finding a constant and a factor to translate one person's utility into another person's. They may even be of quite different shapes. In short, *de gustibus non est disputandam* (There is no disputing about taste).

Does the utility function, as posited by von Neumann and Morgenstern, slope upward? Yes, because a preferred choice indicates more utility (i.e., a larger number). Must the slope decrease? No, not necessarily; von Neumann and Morgenstern do not require this, neither in their axioms nor anywhere else. Hence, for the purposes of game theory, there is no requirement that marginal utility decrease. Risk-taking behavior does not indicate irrationality.

As Christmas of 1942 approached, von Neumann and Morgenstern put the finishing touches on the manuscript. In January 1943, they wrote the preface. The book was done. Now came the time to explain to Princeton University Press that the 100-page pamphlet had grown somewhat—into a 1,200-page manuscript, brimming with graphs and equations. Fortunately, the editors at the press were generous, saying they would do their best to publish the book, even though resources were tight in the midst of World War II. With two five-hundred-dollar grants from Princeton University and from the Institute for Advanced Study, a clean manuscript was retyped. A young mathematician from Japan, at that time a so-called enemy alien, was given the task of entering all formulas from the original manuscript. It went to the printer in 1943, and a yearlong process of typesetting, drafting of diagrams, editing, proofreading, and correcting followed.

Subsidized by an anonymous donation, the 600-page *Theory of Games and Economic Behavior* was printed and published on September 18, 1944. It had been five years since Morgenstern and von Neumann began to collaborate. Morgenstern was definitely the junior partner on the two-man team; he himself admitted that his theoretical abilities were limited. However, it is questionable whether game theory would have developed had the two never met. Morgenstern's crucial contribution was his tendency to ask interesting, provocative questions, thereby serving as a catalyst for von Neumann's genius. True, an intellectual gulf separated him from his senior partner, but that was true of virtually everybody. So, if von Neumann was the father *and* the mother of game theory, Morgenstern was the midwife, as the game theorist Harold Kuhn remarked in the introduction to a commemorative edition of their *magnum opus*. In von Neumann's case, the digression into economics was brief and minor compared to his other contributions to mathematics, computer science, physics, and the atomic bomb.

The book was no overnight bestseller—its break with conventional economics was too great. *Theory of Games* did not concern itself with ordinary maximum or minimum problems, and in terms of real-life economics, it did not limit itself to ordinary exchange or simple monopoly or oligopoly situations. It dealt with exploitation, discrimination, substitution, complementarity, coalitions, power, and the privilege of players; thus, it ventured far beyond economics, reaching into political science and sociology. Initial acceptance of the book was lukewarm at best, and the two authors resigned themselves to the fact that recognition of their theory might have to wait for a generation. As the German physicist Max Planck once remarked, a "new scientific truth does not triumph by convincing its opponents and making them see the light, but rather because its opponents eventually die, and a new generation grows up that is familiar with it."[19] Now, von Neumann and Morgenstern may not have known about that observation, which appeared in print only in Planck's posthumously published autobiography in 1950, but it was very much to the point.

But in a stroke of luck, it took not a generation, but only a year and a half. In March 1946, *The New York Times* published a lengthy review of *Theory of Games* on the front page of its Sunday edition. It began thus: "A new approach to economic analysis that seeks to solve hitherto insoluble problems of business strategy by developing and applying to them a new mathematical theory of games of strategy like poker, chess, and solitaire has caused a sensation among professional economists."[20]

Very flattering—but solitaire? Well, yes, even though there is no opposing player, solitaire is a single-person game of strategy, unlike, say, roulette, which is dominated by probability theory alone and hence presents no interest for game theory. The author of the *Times* piece gave a very cogent overview, partly drawn from a recently published seventeen-page article about the book in the *American Economic Review* by Leonid Hurwicz (winner of the Nobel Prize in Economics in 2007). "One cannot but admire the audacity of vision, the perseverance in details, and the depth of thought displayed on almost every page of the book," Hurwicz wrote, concluding that "the appearance of a book of the caliber of the *Theory of Games* is indeed a rare event." In the April 1946 issue of the *Journal of Political Economy*, Jacob Marschak called it an "exceptional book," and in the *American Journal of Sociology*, Herbert Simon (winner of the Nobel Prize in Economics in 1978) urged every social scientist to master the *Theory of Games*, which seeks to develop "in systematic and rigorous manner a theory of rational human behaviour." Richard Stone (winner of the Nobel Prize in Economics in 1984) called it "a work of immense power and interest" in his sixteen-page review in the *Economic Journal*. The book's impact was also felt in mathematics. In the *Bulletin of the American Mathematical Society*, Arthur Copeland gushed that "posterity may regard this book as one of the major scientific achievements of the first half of the twentieth century."[21]

No wonder *Theory of Games* soon sold out. In 1947, Princeton University Press published the second edition with the appendix that contained the proof of the existence of utility functions. A third edition followed in 1953, and many more came after that. In 2004, a Sixtieth Anniversary Edition was published. Today, Google Scholar lists 35,000 citations to this edition alone, and there are conferences, specialized journals, and literally hundreds of books about game theory. *Theory of Games* must rank up there as one of the defining books of science, arguably on a level with Isaac Newton's *Principia* and Charles Darwin's *On the Origin of Species*.

CHAPTER 8

WOBBLY CURVES

The publication of *Theory of Games* opened the door to a deluge of research that has not abated to this day. Every time one question is answered, a new one arises. One of the first questions that came up was, "Why do people gamble?" Daniel Bernoulli had shown in the seventeenth century that because a person's utility rises at a diminishing rate, he or she is risk averse and therefore willing to pay a premium in order to avoid risks. This is why people insure their homes, cars, and other belongings. The word *premium* is well chosen because the payments demanded by insurance companies are higher than the expected losses. The fact that premiums are over and above the actuarial value of any likely losses is what keeps insurance companies viable and profitable.

But there is another aspect to people's attitude toward risk: Many people like to gamble! This may seem quite surprising indeed. Given the general aversion to risk, who would want to put up money to *assume* risk? It seems counterintuitive.

Actually, it is not all that surprising. Humans have been gambling since time immemorial. Prehistoric evidence of men's propensity toward games of chance includes dice from the third millennium BCE that were unearthed in Mesopotamia (today's Iraq). An Egyptian tablet from about the same time also seems to indicate gambling. The Greek poet Sophocles mentioned dice in a document from the fifth century BCE. And there is evidence that in China, in the third century BCE, lotteries were used to raise money for a war effort, and later to fund the building of the Great Wall.

In spite of their appeal, games of chance were generally deemed disreputable. The Buddha (c.480–400 BCE) denounced games of chance in his Eightfold Path, although a later Indian thinker, the minister Kautilya (371–283 BCE), not only permitted them but even regulated them. He tasked a superintendent of gambling with the administration of gaming activities—in exchange for a 5 percent share of the winnings.

Early Romans allowed gambling until Justinian I (482–565 CE), the Christian emperor of Byzantium, banned it in his fundamental work on jurisprudence, the *Corpus Iuris Civilis*. In fact, most religions cast disapproving eyes on the vice of gambling. Notwithstanding such condemnation, however, many churches do not hesitate to raise money via games of chance such as bingo. Likewise, states sponsor lotteries to fund public works and services like education. In fact, lotteries helped to establish elite institutions like Harvard, Yale, and Princeton.

So we return to the earlier question: Who would put up money to assume risk? Well, there are people who love the adrenaline rush that comes with risky physical activities, like car racing, bungee jumping, rock climbing, off-piste skiing, and the like. But conventional wisdom since Bernoulli's time has it that human beings will not willingly pay to engage in risky monetary ventures. So why would an ordinary John or Jane Doe put up money to buy a lottery ticket? Obviously, the expected payout is lower than the cost of the ticket. After all, it is the premiums over and above expected winnings that keep gambling outfits in business.

What is even more surprising is the fact that there are people who buy insurance against all kinds of risk and—at the very same time—gamble with their hard-earned spending money. What is going on? A person pays money to avoid risk and simultaneously pays out more money to assume it?

Among the first to weigh in on this conundrum were the economist Milton Friedman and his colleague at the University of Chicago, the statistician Leonard Savage. The year was 1948. The thirty-six-year old Friedman (figure 8.1) had spent the last years of World War II at Columbia University, working on problems of weapons design, military tactics, and metallurgy; most important, though, he was a rising star in the world of theoretical economics.

His family hailed from Beregszasz, then in the Hungarian part of the Austro-Hungarian Empire. The town had about 10,000 inhabitants, and

FIGURE 8.1: Milton Friedman.

Source: Wikimedia Commons; the Friedman Foundation for Educational Choice

most of them were Jewish, including the Friedmans.[1] In the late 1890s, two teenagers emigrated separately to the United States, met in New York a few years later, married, and made their home in Brooklyn. This is where Milton was born in 1912. The father was a small-time trader, while his mother worked in a sweatshop. A precocious boy, Friedman entered Rutgers University on a scholarship at age sixteen. He had to augment his meager funds by clerking in a retail store, waiting tables at a restaurant in exchange for free lunch, and working during the summers. Intending to become an actuary, he passed some of the exams and failed others. But then the economics department of the University of Chicago offered a scholarship to do graduate work and that made up his mind.

At Chicago, he had the good fortune to pursue his studies with and under a slew of brilliant economists. He also met a shy and withdrawn, but very bright fellow economics student, Rose Director. Six years later, when their fears of the Great Depression had dissipated, they married. After stints at Columbia, where he completed his PhD, the National Bureau of Economic Research (NBER), the U.S. Treasury Department, and the University of Minnesota, he returned to Chicago. He remained there as the economics department's towering intellect for three decades. Upon his retirement, he continued his research and writing for another three decades at the Hoover Institution at Stanford University, until his death in 2006. In all their years together, Rose would be an active partner with her husband in his professional work.

Friedman was a proponent of monetarism, the theory that money supply influences the national output and can control inflation. He supported free markets, with minimal intervention by government. He advocated freely floating exchange rates, school vouchers, a volunteer army, and the abolition of medical licenses.[2] He was the undisputed leader of the department of economics at Chicago. Like *Austrian Economics*, the moniker *Chicago School* became a mark of distinction. In 1976, Friedman was awarded the Nobel Prize in Economics,[3] and in 1988, he received both the Presidential Medal of Freedom and the National Medal of Science.

Five years younger, Friedman's coauthor of the paper that I shall discuss here was Leonard Savage (figure 8.2), universally known as "Jimmie" (his middle name). His grandparents' family name had been Ogushevitz, which Jimmie's father changed to the more familiar, if slightly ferocious sounding, Savage.

At school, Jimmie did not stand out academically. On the contrary, his teachers thought him feebleminded. But that was due to the boy's very bad eyesight. "He paid no attention to what was going on in school because he couldn't see what was going on in school," his brother recounted in an interview.[4] The fact that the boy was actually brilliant, and most probably bored by the humdrum of his classes, must have exacerbated his teachers' misconception.

In spite of Jimmie's very mediocre grades, his father managed to get him accepted to the University of Michigan to study chemical engineering.

FIGURE 8.2: Leonard Savage.

Source: Leonard Jimmie Savage Papers (MS 695). Manuscripts and Archives, Yale University

Unfortunately, things again went badly. Because of his poor eyesight, he caused a fire in the chemistry laboratory and was expelled. Once more his father intervened, and the young man was allowed to return to the university. But this time the subject was physics, which is, after all, safer than chemistry—at least in a university setting. Eventually, he turned all his attention to mathematics, received his bachelor's degree, got married, and obtained his PhD in pure mathematics three years later.

Recognized as an outstanding postdoc, his next stop was the Institute for Advanced Study at Princeton. He became a statistical assistant to John von Neumann, who recognized his talents and advised him to focus on statistics. Stints at Cornell, Brown, Columbia, New York University, Chicago, Michigan, and finally Yale followed. His tenure at Yale lasted only seven years before he died in 1971, not yet fifty-four years old.

Savage's most noted work was the book *Foundations of Statistics*, published in 1954. Influenced by Frank Ramsey's groundbreaking work (see chapter 6) and von Neumann and Morgenstern's game theory (see chapter 7), he proposed a theory not only of subjective utility but also of personal probability, based on one's degree of conviction. The theory was based on a set of axioms, one of which was the apparently innocuous "Independence of Irrelevant Alternatives" axiom of von Neumann and Oskar Morgenstern. As we shall see in chapter 10, this axiom cast a dark cloud over the entire theory of expected utility.

It was at the University of Chicago that Friedman, the professor of economics, and Savage, then a research associate in statistics, collaborated on a paper that would become famous. Just a year earlier, the second edition of *Theory of Games* had been published, with the all-important proof that adherence to the von Neumann-Morgenstern axioms is a necessary and sufficient condition for a person to possess a utility function.

We know that whenever a decision-maker has to choose among several riskless alternatives, he will choose the one that maximizes the payout,[5] which in this case is the same thing as maximizing utility. But in the presence of riskiness, it is the *expected utility of the payout* that is the factor to be maximized . . . which is not the same thing as the *utility of the expected payout*. Because marginal utility of money diminishes as wealth increases, maximizing the expected utility is no longer equivalent to maximizing the expected payout. And because marginal utility of money diminishes for everybody, and diminishing marginal utility is equivalent to risk aversion, it is obvious that everybody must avoid risk. Or is it?

Despite the justifications that von Neumann and Morgenstern provided for this theory, which had been accepted as plausible since Bernoulli's time, Friedman and Savage were perplexed by the profound paradox to which I alluded at the chapter's outset: Why do so many risk-averse people, who buy insurance against all kinds of risks, also engage in gambling activities? Why do homeowners who insure their property also buy lottery tickets? "Offhand, it seems inconsistent for the same person both to buy insurance and to gamble: he is willing to pay a premium, in the one case, to avoid risk, in the other, to bear risk," they wrote. But the paradoxical phenomenon is by no means rare, they remarked; rather, it is so pervasive that "[m]any governments find . . . lotteries [to be] an effective means to raise revenue."[6]

After reviewing all kinds of behavior that reveal human beings' propensity for taking risks, they concluded that "it turns out that these empirical observations are entirely consistent with the [von Neumann-Morgenstern] hypothesis if a rather special shape is given to the total utility curve of money." Now, what is this "rather special shape"?

So far, we have repeated again and again, both in this chapter and in the previous ones, that marginal utility for wealth decreases as wealth increases: Utility for wealth rises, but does so less and less as wealth increases. (In other words, the utility curve's slope becomes flatter and flatter.) After all, as already stipulated, a second scoop of ice cream provides less pleasure than the first, and an additional dollar provides less utility to a millionaire than to a pauper. Depicted on graph paper, the curve of the utility function rises, but it bends downward or, in technical terms, the shape of the utility function is concave when viewed from below. But maybe this is not so throughout the wealth spectrum? Maybe there are pockets of wealth, somewhere between one dollar and a million dollars, where the marginal utility for wealth increases?

This is exactly what Friedman and Savage claimed. For example, they argued, an additional chunk of money may allow a member of the working poor to jump into the middle class. Now, that would be really worthwhile. This chunk may change his life. Hence, while one additional dollar may be all but unnoticeable, 10,000 additional dollars may offer more than 10,000 times the utility of a single additional dollar. The implication is that in this region of wealth, marginal utility increases. And that, of course, implies that the individual would be willing to engage in a gamble even if the odds were to his disadvantage. (By the way, recall in this context that von Neumann and Morgenstern did not stipulate that marginal utility must decrease, i.e., that the slope of the utility curve must become flatter.)

To illustrate, let's take Sheila, a moderately well-off woman with, say, a home worth $100,000 and $50,000 in her bank account. She will probably have insurance for her house because between zero and about $150,000, the function depicting her utility for wealth is convex. But what she really wishes for is to enter the "quarter-million dollars and above" bracket. To get a shot at attaining that objective, Sheila is willing to pay, say, ten dollars to buy a lottery ticket that would give her a chance of winning $100,000 . . . even though she knows that she will probably lose her investment. Her willingness to engage in the gamble implies that her utility function is convex between about $150,000 and $250,000.

If, against all odds, she does win the lottery and becomes a quarter-millionaire, she will protect her newfound status by being risk averse again. Hence, beyond a quarter of a million dollars, her utility function is again concave (see figure 8.3). What all this means is that at her current wealth, $150,000, she is willing both to purchase lottery tickets and to buy insurance for her home. Paradox resolved!

So, a wiggly utility function—which can be given a "tolerably satisfactory interpretation," the paper's authors say—explains why many individuals may simultaneously buy insurance and gamble. The wiggles may depict different socioeconomic classes, with the initial and terminal concave segments

FIGURE 8.3: The utility function of Sheila's wealth.

designating low and high wealth, while the convex segment between them indicates a willingness to take risks in order to jump into the higher class. Moreover, wiggliness is quite consistent with the von Neumann-Morgenstern framework: None of the axioms are violated, and decision-makers still maximize utility—albeit a utility whose graph is convex in certain regions.

The theory that Friedman and Savage put forth explains all sorts of risky behaviors, not just lottery gambling and betting in casinos; investment decisions, occupational choices, and entrepreneurial undertakings are also covered. One riddle continued to puzzle them, however: "Is it not patently unrealistic," they asked themselves, "to suppose that individuals consult a wiggly utility curve before gambling or buying insurance, that they know the odds involved in the gambles and insurance plans open to them, that they can compute the expected utility of a gamble or insurance plan, and that they base their decision on the size of the expected utility?"

Yes, that would be patently unrealistic; surely, nobody inspects his or her utility function and performs complex calculations before making a decision. However, the objection is not relevant, according to Friedman and Savage. What is relevant, they claim, is that decision-makers behave *as if* they inspected their utility functions, *as if* they knew the odds, and *as if* they calculated the expected utilities. The validity of a theory depends solely on whether it yields sufficiently accurate predictions about the class of decisions with which the hypothesis deals. It is, they say, as if a billiard player knew the equations of elastic collisions, could estimate accurately by eye the angles, make lightning calculations, and then perform the shot. The proof of the pudding is in the eating. And on that count, wiggly utility curves do very well.

Not long after the paper was published, a perceptive graduate student noted that Friedman and Savage, in their attempt to explain the insurance-gambling paradox, had inadvertently introduced a different paradox. Their model implied behavior that would, if it truly occurred in real life, be counterintuitive. Take Alberta, a person of medium wealth, exactly halfway between poverty and affluence. According to Friedman and Savage, the utility function is concave for low wealth and for high wealth, and convex in between. Alberta finds herself in the middle of the convex (i.e., in the risk-taking) region of wealth.

Now, present Alberta with the following gamble: At the flip of a coin, the loser becomes poor and the winner affluent. The gamble is fair in actuarial terms because the expected payout is zero: If the gamble were repeated many times, Alberta would, on average, remain at her current wealth.

In the Friedman-Savage framework, Alberta, willing to take risks at her current wealth, should jump at the occasion. But in practice, no sane person of medium wealth would ever participate in a fair gamble that would render her either rich or poor, claimed the student. Why should she? Just to end up, on average, where she started? But in Friedman and Savage's world, Alberta would love such gambles. So, the theory needed further developing.

The graduate student's name was Harry Markowitz. Born in 1927 in Chicago, the only child of Morris and Mildred Markowitz, he grew up during the Great Depression. His parents owned a small grocery store and, being in the food and dry goods business, were not severely affected by the general despair that had gripped the country. There was always enough to eat, and the boy had his own room. He enjoyed baseball and tag football, played the violin in the high school orchestra, and read a lot.

At college, he was inclined toward philosophy and physics. But upon graduation after two years at the University of Chicago with a bachelor's degree in the liberal arts, he decided on economics. He was interested in microeconomics and macroeconomics, but what really fascinated him were the economics of uncertainty. He read *Theory of Games* and became engrossed with von Neumann and Morgenstern's arguments concerning expected utility, Friedman and Savage's utility function, and Savage's personal utility ideas. At Chicago, he had the good fortune of counting Friedman and Savage among his teachers.

Once he had become aware of the inconsistency in the Friedman-Savage paper, Markowitz felt the need to dig further. He did a survey among his friends, asking them the following questions:

- Do you prefer to get 10 cents for sure, or one chance in ten of getting $1?
- Do you prefer to get $1 for sure, or one chance in ten of getting $10?
- Do you prefer to get $10 for sure, or one chance in ten of getting $100?
- Do you prefer to get $100 for sure, or one chance in ten of getting $1,000?
- Do you prefer to get $1 million for sure, or one chance in ten of getting $10 million?

"The typical answers (of my middle-income acquaintances) to these questions are as follows," he wrote: Most of them preferred to take their chances and gamble for the dollar rather than getting a dime for sure. Most also preferred to gamble for $10 rather than getting $1 for sure. But then preferences began to differ. Some preferred $10 for sure rather than a gamble for $100, but others preferred it the other way around. He also found differences of opinion in the $100 dollars for sure versus a gamble for $1,000. But when he reached the last option—a million for sure, or a gamble for 10 million—all his friends, without exception, preferred getting a million for sure.[7]

Markowitz then asked the questions in the other direction:

- Do you prefer to owe 10 cents for sure, or one chance in ten of owing $1?
- Do you prefer to owe $1 for sure, or one chance in ten of owing $10?
- Do you prefer to owe $10 for sure, or one chance in ten of owing 100?
- Do you prefer to owe $100 for sure, or one chance in ten of owing 1,000?
- Do you prefer to owe $1 million for sure, or one chance in ten of owing $10 million?

In general, people preferred to owe 10 cents for sure, rather than one chance in ten of owing a dollar, and owe one dollar for sure rather than one chance in ten of owing $10. Thereafter, opinions differed again, and when he got to the biggie, "the individual generally will prefer one chance in ten of owing $10,000,000 rather than owing $1,000,000 for sure."[8]

What Markowitz's survey demonstrated was that his friends behaved differently depending on whether they were about to get something or about to owe something. When confronted with potential receipts, they take risks for small amounts but are risk averse for large amounts. When confronted with potential losses, though, they avoid risks for small amounts but take risks for large amounts.

What does that mean for the shape of the utility curve? Friedman and Savage had posited three regions in their paper—concave for low wealth, convex for medium weath, and concave again for large wealth. Markowitz added a further wobble—he fixed the person's current wealth in the middle, at the cusp between the concave and the convex regions (see figure 8.4). To the right of the current wealth, for gains, the utility curve is, at first, convex. To the left,

FIGURE 8.4: Friedman's wobbles, Markowitz's wobbles.

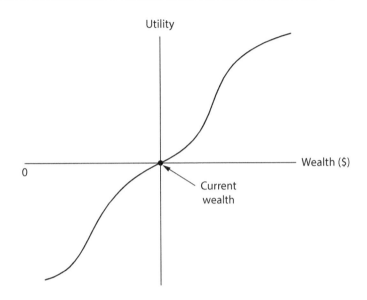

for losses, it is, at first, concave. But farther out, there are two more regions: For very large gains, the curve becomes concave again, and for very large losses, it becomes convex.

These facts were borne out, Markowitz claimed, by "a common observation that, in card games, dice games, and the like, people play more conservatively [i.e., are risk averse] when losing moderately, more liberally [i.e., will take more risk] when winning moderately."

Markowitz made several additional comments about how utility curves should be shaped—they fall faster to the left (for losses) than to the right (for gains), they are bounded from above and from below, and the final wobbles are further removed from current wealth for the rich than for the poor. Like Friedman and Savage's utility function, Markowitz's multiwobbled shape explains why people may engage simultaneously in insurance and in gambling. But, in contrast to Friedman and Savage, it avoids the paradox of Alberta accepting an actuarially fair gamble that would make her either poor or rich.

According to modern standards, Markowitz's paper is not very rigorous. The lack of real-life verification of his propositions—apart from a cursory poll among his friends—made it a dubious candidate for publication.[9] Apparently, Markowitz himself was aware of that, as he revealed in the paper's closing paragraph: "It may be objected that the arguments in this paper are based on flimsy evidence . . . I realize I have not demonstrated 'beyond a shadow of a doubt' the 'truth' of the hypotheses introduced. I have tried to present, motivate, and to a certain extent, justify and make plausible a hypothesis which should be kept in mind when explaining phenomena or designing experiments concerning behavior under risk or uncertainty." As justifications go, such a message to referees and readers should have raised red flags. Nevertheless, utility functions shaped like Friedman and Savage suggested, or like Markowitz proposed, do seem to be the answer to the insuring/gambling paradox.

In any case, the "Utility of Wealth" paper was only a sideline for Markowitz. The paper that truly and justifiably made him famous, "Portfolio Selection," was published the same year in the *Journal of Finance*.[10] While still only a doctoral student, Markowitz decided to apply mathematical methods to the stock market. The idea had come up during a chance conversation with colleagues. According to conventional wisdom at the time, the value of a stock is the present, discounted value of its future dividends. But because future dividends are uncertain, Markowitz interpreted this to mean that the stock's value is determined by the present, discounted value of its *expected* future dividends. "But if the investor were only interested in expected values of securities," Markowitz pointed out, "he or she would only be interested in the expected value of the portfolio; and to maximize the expected value of a portfolio, one need invest only in a single security. This, I knew, was not the way investors did or should act. Investors diversify because they are concerned with risk as well as return."[11]

FIGURE 8.5: Harry Markowitz receiving the Nobel Prize.

Courtesy of the Harry Markowitz Company

Thus was born modern portfolio theory. The qualifier *modern* is required here because we saw in Chapter 1 that the idea of diversification had already been proposed three centuries earlier by Daniel Bernoulli. For this paper, and for his many subsequent contributions to the development of financial economics, Markowitz was awarded the Nobel Prize in Economics in 1990 (figure 8.5).

CHAPTER 9

COMPARING THE INCOMPARABLE

As pointed out previously, a weakness of utility theory is that utility cannot be compared between people. The problem is measurement. Counting units is the most direct way to measure something. The crowd size at Donald Trump's inauguration was definitely smaller than at either of Barack Obama's, "fake news" tweets notwithstanding. Weights and lengths also can be objectively measured and compared. The prototype of the kilogram, the *Kilogramme des Archives*, and the prototype of the meter bar, the *Mètre des Archives*, both stored at the International Bureau of Weights and Measures near Paris, served as the standards for all kinds of measurements, and the substance being measured is immaterial.[1] Hence, 2 kilograms of potatoes weigh twice as much as 1 kilogram of carrots. A 20-centimeter wooden ruler is one-third as long as a 60-centimeter metal rod. The problems begin when artificial scales are used.

We already pointed out in Chapter 6 that the Mohs scale measures the relative hardness and resistance to scratching between minerals. Hence, diamonds (hardness 10) can scratch quartz (hardness 7), and talc (hardness 1) can be scratched by any other mineral. However, because it only measures relative hardness, the Mohs scale is only an ordering. Hence, topaz (hardness 8) is harder than fluorite (hardness 4), but this does not mean that it is twice as hard.

With temperature, one can at least provide an objective measurement. But even then, we cannot say, for example, that Phoenix, at 100°F, is twice as hot as Boston, at 50°F. This becomes immediately obvious when we translate the temperatures into degrees Celsius: Phoenix at 37.4°C and Boston at 9.9°C. In both measuring systems, Phoenix is hotter than Boston. But twice as hot? No way!

Utility is in a different league altogether. Not only can it not be objectively measured like crowd size, weights, time spans, and lengths, it cannot even be compared like the hardness of minerals. As a psychological concept, it resists measurement as we know it. Emotional sensations are totally subjective and cannot be objectively gauged. Take pain, for example. Does a woman at childbirth experience more distress than a man suffering from a kidney stone? Or pleasure—How does the joy of seeing one's boisterous grandchildren compare to the enjoyment of a peaceful walk in the park?

Likewise, utility cannot be measured numerically. The functions and graphs that describe a person's utility, as postulated by John von Neumann and Oskar Morgenstern—and before that, by Daniel Bernoulli and Gabriel Cramer—are defined as orderings that are relevant only for that particular person. In fact, one may use all kinds of mathematical manipulations on utilities, so long as they remain *monotone*, which means that they always rise. For example, utilities could be squared, or multiplied by 10, without changing the ordering. In particular, the utility functions that von Neumann and Morgenstern postulated can be linearly transformed (i.e., a utility can be multiplied by a constant and then another constant may be added without changing the ordering). Note that degrees Celsius are just a linear transformation of degrees Fahrenheit.[2] Hence, if it is hotter in Celsius, it is also hotter in Fahrenheit, even though one may not compare the numerical values directly.

Recall that Bernoulli suggested the logarithm as an indicator of the utility of wealth, while Cramer proposed the square root. Both are legitimate indicators, but while each person's decisions are accurately described by his or her own utility function, interpersonal comparisons—like comparisons of degrees Fahrenheit and Celsius—are not permissible. One person's utility for an orange cannot be compared to another's utility for a banana. One cannot even compare one person's utility for grapes with another's utility for the same grapes, nor can one person's utility for the first cookie be numerically compared to her or his utility for the tenth, except to say that the tenth adds less utility than the first.

The reason for the impossibility of interpersonal comparisons of utility is that utility, like the Mohs scale, is no more than an ordinal measure (i.e., a ranking), rather than being a cardinal measure that allows addition and subtraction, like kilometers and centimeters. Hence, one can state that 1,000 dollars gives Sam more utility than 500 dollars, but not how much more. And one certainly cannot assert that Sam derives more utility, or less utility, from 200 dollars than does Tom.

Surprisingly, however, there is one thing that *can* be compared between people: their degree of risk aversion. Two professors—one on the West Coast, the other on the East—developed a measure to gauge people's attitude toward risk, all without being aware of each other's work. They were Kenneth Arrow, a professor of economics at Stanford, and John Pratt, a professor of business administration at Harvard.

Kenneth Joseph Arrow (1921–2017, figure 9.1) was born in New York City and spent his youth and student days in the Big Apple.[3] The family lived in comfortable circumstances until the Great Depression wiped out most of their wealth; for the next ten years, they lived in poverty. When the time came for Arrow to attend college, his parents could barely afford the cost. Fortunately, the City College of New York (CCNY) offered city residents a higher education without tuition fees, and Arrow was forever grateful for the opportunity accorded him. At CCNY, he majored in mathematics with minors in history, economics, and education—and had the intention of becoming a math teacher. But when he graduated, winning the Gold Pell Medal for the highest grades, there were no positions available in the New York City school system. So he entered Columbia University to continue studying mathematics. He received his master's degree in 1941 and then did not really know what to do next.

To his great fortune, he had taken a course at Columbia in mathematical economics with Harold Hotelling, a statistician who held an appointment at the department of economics. This experience proved to be decisive: Arrow decided that mathematical economics was the subject to which he would henceforth devote his life. A fellowship to the department of economics ensued, but then life was interrupted by World War II. He entered the U.S. Air Corps in 1942 as a weather officer, rising to the rank of captain in the Long-Range Forecasting Group. One day, using their academic training,

FIGURE 9.1: Kenneth J. Arrow.

© Chuck Painter / Stanford News Service

Arrow and his colleagues decided to submit their work to a statistical test. They investigated whether the group's aim—forecasting the number of rainy days one month in advance—was being achieved. Not surprisingly, the conclusion was that it was not. In a letter to the general of the Air Force, they advised the dissolution of the Long-Range Forecasting Group. The response came half a year later: "The general is well aware that your forecasts are no good. However, they are required for planning purposes."[4] So the group continued to prognosticate sunny days and rainy days using techniques that were about as good as drawing lots from a hat. Arrow left the Air Force in 1946. Something positive nevertheless originated from Arrow' s work with the military: His first scientific paper, entitled "On the Optimal Use of Winds for Flight Planning," was published in the *Journal of Meteorology* in 1949.

After the war, Arrow continued his graduate work at Columbia. Mindful of the hardships that his family had suffered during the Depression, he was on the lookout for a solid, down-to-earth profession. For a while, he toyed with the idea of becoming a life insurance actuary—and actually passed a series of actuarial exams. However, as he searched for a job in the insurance industry, an older colleague dissuaded him, and Arrow decided to embark on a career in research. In 1947, he joined the Cowles Foundation for Research in Economics at the University of Chicago. There, he encountered "a brilliant intellectual atmosphere... with eager young econometricians and mathematically inclined economists." It was also there that he met Selma Schweitzer, a young graduate student, whom he subsequently married. She was at the Cowles Foundation on a fellowship designed for women pursuing quantitative work in the social sciences. (Originally, the fellowship indicated preference for "women of the Episcopal Church," but the religious affiliation was subsequently dropped—which was fortunate, because Selma, like Arrow, was Jewish.)

Starting in 1948, a few years after von Neumann and Morgenstern had published their *Theory of Games and Economic Behavior,* Arrow spent summers at the RAND Corporation in Santa Monica, California, the original nonprofit, global policy research institute that would set the standard for all think tanks that followed. In 1949, Arrow was appointed acting assistant professor of economics and statistics at Stanford University and then advanced through the ranks, eventually becoming professor of economics and professor of operations research. Except for an eleven-year interlude at Harvard University and visits to Cambridge, Oxford, Siena, and Vienna, Arrow spent his whole career at Stanford until he retired in 1991. Among the numerous prizes he received were the John Bates Clark Medal in 1957, awarded every year to an outstanding economist under forty, and of course, the Nobel Prize in Economics, which he won with John Hicks in 1972. Elected to the National Academy of Sciences and to the American Philosophical Society, he has received more than twenty honorary degrees. Even the Vatican weighed in, making him a member of the Pontifical Academy of Social Sciences. During his rich career, Arrow also served as president of the Econometric Society and was on the staff of the U.S. Council of Economic Advisors and a fellow and a member of numerous learned societies. He died in 2017.

One of Arrow's most famous accomplishments was actually contained in his doctoral thesis: the illustrious Impossibility Theorem, which states, "If we exclude the possibility of interpersonal comparisons of utility, then the only methods of passing from individual tastes to social preferences which will be satisfactory and which will be defined for a wide range of sets of individual orderings are either imposed or dictatorial."[5] In other words, there exists no reasonable method to elect a leader democratically. Even the touted majority rule has a very serious shortcoming.[6] And what is the reason for this depressing state of affairs? The impossibility of interpersonal comparison of utility!

In the early 1960s, while Arrow lectured at Stanford on economics, particularly the economics of uncertainty, a young professor across the country was busy researching similar topics at the Harvard Business School. John Winsor Pratt, born in 1931, was educated at Princeton and obtained his PhD in 1956 in statistics at Stanford University. His thesis, "Some Results in the Decision Theory of One-Parameter Multivariate Polya-Type Distribution," presaged his lifelong interest in decision-making. In 1962, he was elected as Fellow of the American Statistical Association, and in 1988 as a member of the American Academy of Arts and Sciences. He coauthored a book entitled *Introduction to Statistical Decision Theory*, published by MIT Press in 1995, and chaired National Academy of Sciences committees on environmental monitoring, census methodology, and the future of statistics. In recognition of his contributions to the field of decision analysis, Pratt was awarded the Frank P. Ramsey Medal in 1999 by the Decision Analysis Society. The prize, the highest award that the society gives, was established in memory of the Cambridge mathematician who is the subject of Chapter 6.

One paper in particular made Pratt famous. Entitled *Risk Aversion in the Small and in the Large*, it was published in *Econometrica* in January 1964 and became one of the most frequently cited economics papers of its time. In recent years, it is no longer quoted as often, not because it has lost its relevancy, but rather because the concept that Pratt discussed has entered the mainstream and no longer needs to be mentioned at every turn.

"This paper concerns utility functions for money," Pratt began. The paper discussed a "measure of risk aversion in the small, the risk premium or insurance premium for an arbitrary risk."[7] With this, he let the cat out of the bag: While utility can only be ranked, not measured in the traditional sense, Pratt asserted that the degree of a person's risk aversion could be measured by relating it to the insurance premiums that individuals are willing to pay to avoid risk. The approach is similar to Frank Ramsey's, who related the degree of belief in a probability to the distance that one would be willing to walk out of one's way (see Chapter 6).

Before we describe what measure Pratt proposed to gauge risk aversion, let me recap Arrow's work in this endeavor. In 1962, about the time that Pratt was working on the development of his measure of risk aversion, Arrow taught a course called "Economics of Uncertainty" to his Stanford students. Lecture 6 dealt with liquidity preference (i.e., the preference that people have for riskless cash, even though it carries no interest, as compared to risky investments that are expected to produce a return, but also may produce a loss). In particular, Arrow discussed the optimal amount of money that a decision-maker should invest in a speculative venture, while keeping the remainder of her or his wealth in cash. He showed that this amount depends on how wealthy the investor is. In the course of his derivation, Arrow introduced a mathematical expression that measures the investor's degree of risk aversion. It was the exact same expression that Pratt had also found.

As has been noted, Pratt was unaware of Arrow's work. This was before the Internet, and practically nobody outside the circle of Arrow's Stanford students had ever heard of the measure. Shortly thereafter, Arrow received an invitation to deliver lectures at the first Yrjö Jahnsson Lectures in Helsinki. This lecture series, named after a Finnish economics professor who had made a fortune in business but died relatively young, has become justly famous—ten of the lecturers were later awarded the Nobel Prize in Economics. Arrow decided to present his work on risk aversion. He delivered the lectures in December 1963 under the title "Aspects of the Theory of Risk Bearing." It was only then that Pratt learned about Arrow's findings. With his own paper about to be published, he managed to add a section at the end, entitled "Related Work of Arrow." The mathematical expression that the two professors had found independently of each other would henceforth become known as the Arrow-Pratt measure of risk aversion. So, while utilities cannot be compared between individuals, it turns out that risk premiums can be.

As you may recall from Chapter 1, a person's risk premium is determined by the horizontal stretch between the utility curve and the straight line that slopes from one wealth level to the other. To save you, the reader, the trouble of paging back, I will repeat the argument and the graph from that chapter (see figure 9.2).

Let's say that Monsieur Pimpodou faces a 50–50 chance of ending up with either 2,000 or 3,000 ducats. His expected wealth, therefore, is 2,500 ducats (50 percent of 2,000, plus 50 percent of 3,000). Being risk averse, he would settle for 2,440 ducats if he were guaranteed that sum for sure. The mathematical explanation went like this:

> First, we indicate a wealth of 2,000 ducats on the $-axis ($a$). We trace this value to the utility curve (b), which indicates a utility on the U-axis of 7.6 (c). We do the same for 3,000 ducats, on the $-axis ($d$), tracing it to the utility curve (e), which indicates a utility of 8.0 on the U-axis (f). Now we seek the expected utility on the U-axis, which is halfway between 7.6 and 8.0 [i.e., at 7.8 (g)]. (It is halfway because the odds are 50–50. If the odds were different, the location on the U-axis would have to be adapted accordingly.) The question is: What ducat value corresponds to a utility of 7.8? Let us trace from the expected utility level of 7.8 toward the utility function (h), which indicates a value of 2,440 on the $-axis ($i$). However, the mathematically expected monetary value of the gamble lies halfway between 2,000 and 3,000 ducats [i.e., at 2,500 ducats (j)]. Now we are done. A final, certain wealth value of 2,440 ducats gives Monsieur Pimpodou the same utility that he would expect to enjoy from gambling on the proposal. The monetary difference between i and j, 60 ducats, is the price that he is prepared to pay to avoid the gamble.

Another person in the same situation may be willing to forgo 70 ducats, or 55, or some other amount. In contrast to their utilities, the risk premiums that

FIGURE 9.2: The utility of wealth.

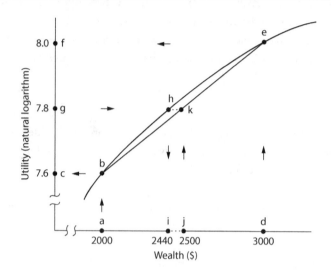

Note: the curve (through points *b*, *h*, and *e*) is the natural logarithm; that through points *b*, *k*, and *e* is simply a straight line.

they are willing to forgo *can* be compared. Hence, the decision-makers' degrees of risk aversion are determined by the length of the horizontal strips.

The crucial question now is: How does the graph of the utility function govern the length of this strip? Is it the slope of the curve, $U'(wealth)$? Or its curvature, $U''(wealth)$? The answer to both questions is no, because different slopes and different curvatures may result in the same risk premium.[8] Hence, neither the slope nor the curvature alone determines the degree of risk aversion. What Arrow and Pratt found was a measure of risk aversion, as a function of wealth, that combines both the utility's slope *and* its curvature:

Risk aversion as a function of wealth $= -U''(wealth)/U'(wealth)$.

In effect, Arrow and Pratt defined the measure of risk aversion as the utility function's curvature normalized by its slope. (In other words: How many times greater than the slope is the curvature?) Because U'' is negative for risk averters while U' is positive for everybody, the degree of risk aversion is generally a positive number.[9] In addition, Arrow and Pratt defined another measure—namely, the degree of *relative risk aversion*, which is the absolute risk aversion, as defined here, multiplied by wealth.

An interesting question is whether risk aversion increases, decreases, or stays constant as wealth increases. Because people are different, there is no correct answer. Some may become less risk averse as they grow richer, but others may not. The implications are important. As Pratt pointed out in his paper, decreasing absolute risk aversion means that the richer he is, the less the decision-maker is willing to pay to insure against a certain risk, and vice versa. Or, as Arrow put it in his lectures, decreasing absolute risk aversion implies that the richer she becomes, the more will the investor invest in risky ventures. The same holds for relative risk aversion. Decreasing relative risk aversion means that an investor will invest a greater proportion of her wealth in risky ventures, or else be willing to pay only smaller premiums against risks that threaten a proportion of his assets.

Numerous studies have determined that people usually exhibit absolute risk aversion that decreases with wealth, and that their relative risk aversion generally hovers around 1.0.[10] This is interesting because it implies, after some mathematical manipulation, that most people's utilities are best described by the logarithmic function, as Bernoulli asserted back in 1738. And this, in turn, implies that no matter how rich they are, investors generally invest a constant proportion of their wealth in risky ventures.[11]

PART THREE

... BUT MAN IS THE MEASURE OF ALL THINGS

CHAPTER 10

MORE PARADOXES

Maurice Allais was barely three years old when his father was called to military service in August 1914 to fight in World War I. The little French boy would never see him again; captured by the Germans, he died in captivity seven months later. The loss marked the boy not only in his youth, but for his entire life.

Born into a typical French working-class family, Allais' maternal grandfather had been a carpenter, the father a cheesemonger with a little shop in Paris. After his father's death, his mother raised Maurice while hovering on the brink of poverty. The boy showed early signs of exceptional gifts. He excelled in high school, usually being the first in his class in almost all subjects, be they French, Latin, or mathematics. But it was history that fascinated him. Allais first intended to study that subject upon graduation but was dissuaded by his math teacher, who convinced him that with talents such as his he should turn to science. The French system of higher education is sometimes decried as elitist, but it is a tribute to it that gifted young people need not be well-to-do or have the right contacts to enter the leading institutions. Allais gained entry to the elite and highly competitive *École Polytechnique* in 1931 and graduated first in his class two years later.

Upon graduation, the young man visited the United States. It was 1933, and that country was in the midst of the Great Depression. Aghast at the devastation he saw, he decided that he needed to understand how such events could occur. It was the sight of "a graveyard of factories" that led him to study economics. "My motivation was an idea of being able to improve the conditions of life, to try to find a remedy to many of the problems facing the world. That's what led me into economics. I saw it as a way of helping people."[1]

Back in France, high-level positions in governmental administration beckoned, but like all graduates of *École Polytechnique*, Allais had to complete a year of military service first. He did so in the artillery corps and then completed two further years of study toward an engineering degree at the *École Nationale Superieure des Mines* in Paris. In 1937, the newly minted, twenty-six-year-old mining engineer was put in charge of mines and quarries, as well as the railway system, in five of France's eighty-nine *départements*. But soon World War II was in full swing, and Allais was called back to military service. As a lieutenant, he was given command of a heavy artillery battery on the Italian front, but after the fall of Paris to the Germans, the war with Italy only lasted for two weeks, from June 10 to June 25, 1940. Following the armistice, Allais was demobilized and returned to his former position in German-occupied France.

It was during the last years of the war, and in the three years following it, that Allais formulated his thinking about the economy, all the while carrying out his administrative functions. Largely self-taught, and working eighty hours a week, Allais published his first two fundamental works, *À la Recherche d'une Discipline Économique* (In quest of an economic discipline, 1943) and *Économie et Interêt* (Economy and interest, 1947), as well as three minor pieces and various news articles. He had decided to write and publish all his works in French, his native tongue—a choice that would prove fateful for his career.

In 1948, he was excused from administrative duties and henceforth concentrated exclusively on research, teaching, and scientific publication. In addition to being a professor of economic analysis at the *École Nationale Supérieure des Mines* from 1944 on, and the director of a research unit at the *Centre de la Recherche Scientifique* (CNRS) from 1946 on, he held teaching positions at other institutions, such as the institute of statistics at the University of Paris (1947–1968), the Thomas Jefferson Center at the University of Virginia (1958–1959), the Graduate Institute of International Studies in Geneva (1967–1970), and again the University of Paris (1970–1985).

His many contributions to economic science derived essentially from the epiphany he had had during his visit to the United States as a student. He sought to find solutions to the fundamental problem of any economy: how to maximize economic efficiency while ensuring an acceptable distribution of income. "Thus, my vocation as an economist was not determined by my education, but by circumstances. Its purpose was to endeavour to lay the foundations on which an economic and social policy could be validly built."[2]

The numerous honors for his scientific work were crowned with the award of the Nobel Prize in Economics in 1988, for his "pioneering contributions to the theory of markets and efficient utilisation of resources." Allais was seventy-seven years old, and many observers wondered why it had taken the Nobel committee so long to honor this French economist, especially since other winners—Paul Samuelson in 1970, John Hicks and Kenneth Arrow in 1972, and Robert Solow in 1987—had been awarded the prize for work that Allais had already done, or for which he had laid the groundwork. Even a *student* of his, Gérard Debreu, had received the Nobel Prize five years before. Why was that? The reason was simple: While the lingua franca in economics was English, as in all sciences, Allais had written his work in French. Had his early publications been available in English, "a generation of economic theory would have taken a different course," according to Paul Samuelson.[3]

Allais's interests also included applied economics (namely, economic management, taxation, the distribution of income, monetary policy, energy, transportation, and mining). Motivated by his studies of the factors of economic development, the liberalization of international trade, and the monetary conditions of international economic relations, he took an active part in various organizations, such as the European Union of Federalists, the European Movement, the Movement for an Atlantic Union, and the European Economic Community. He was a rapporteur at many international conferences aimed at establishing the North Atlantic Treaty Organization (NATO) and the European Community.

He was opposed, however, to the creation of a common European currency—not because he disapproved of the euro as such, but because he thought such a step must be preceded by a full political union of the European nations. And even though he was one of the founding members of the prestigious liberal think tank *Société du Mont Pélerin*, he refused to sign the constituting document because in his opinion, too much weight was given to private property rights.

Surprisingly, economics was not the only scientific field in which Allais made his mark. Trained as a scientist, he remained fascinated by physics and performed experiments in the early summer of 1954 that produced baffling results. Over a period of thirty days, he recorded the movement of a specially designed "paraconical pendulum." While the measurements were in progress, a total solar eclipse of the Sun occurred, and at the very moment that the Moon passed in front of the Sun, the pendulum sped up.

This result, which has since been reproduced during about twenty solar eclipses, was extremely puzzling and remains unexplained to this day. One possible explanation is that the Moon, when it passes in front of the Sun, either absorbs or bends gravitational waves. That would imply that space manifests different properties along different axes due to motion through an ether that is partially influenced by planetary bodies. However, the theory of an ether has been debunked since 1887. Or has it? Did Allais's experiment revive the discredited theory, thereby also putting into doubt parts of Einstein's theory of relativity? The latter would have pleased Allais, who was convinced that Einstein had plagiarized the work of Henri Poincaré.[4] Had Allais also found a convincing explanation for what would become known as the *Allais effect*, it would perhaps have warranted another Nobel Prize, this time in physics.[5]

It is one of Allais's early works, dating back to 1952, that is of interest here. In previous works, Allais had proved that equilibrium in a market economy is equivalent to maximum efficiency. Now he wanted to extend that theory to an economy with uncertainty.

John von Neumann and Oskar Morgenstern had published the second edition of *Theory of Games* five years earlier, including the proof that adherence to their four axioms was a necessary and sufficient condition for a person to possess a utility function. And according to them, any rational operator must maximize the mathematical expectation of his or her utility.

To Allais, this stance was unacceptable because it neglected the fact that human decision-makers were . . . well, human. He concluded that *mathematical* expectation could not be the central factor when making decisions; the crucial element had to be *psychological*. It is the probability distribution of psychological values around their mean, not the probability distribution of utilities around their mean, that represents the fundamental psychological element of the theory of risk.

In stipulating a utility function that increases at a decreasing rate, Daniel Bernoulli had already taken into account human psychology to a large degree. Later on, von Neumann and Morgenstern, as well as Milton Friedman, Leonard Savage, and Harry Markowitz, had advanced this theory, and Frank Ramsey and Savage had postulated that even probability—an objective measure if ever there was one—had to be considered from a psychological perspective. But Allais took an immense step further.

To test the underlying assumptions of the theory of expected utility, particularly its axioms, he devised an intricate experiment with about 100 subjects, all of them well versed in the theory of probability. By any reasonable standard, they would be considered to behave rationally. Allais asked them two questions. While reading them, please ask yourself which answers you would give. The first question was:

Which of the following situations do you prefer?[6]

(a) $1 million for sure,

or

(b) 10 percent chance of receiving $5 million,
    89 percent chance of getting $1 million, and a
    1 percent chance of getting nothing.

Which of the two situations would you, the reader, prefer?

Most of Allais's subjects answered that they preferred (*a*), the $1 million for sure. Then Allais asked the second question:

Which lottery do you prefer:

(c) 11 percent chance of getting $1 million, and
    89 percent chance of getting nothing,

or

(d) 10 percent chance of getting $5 million, and
    90 percent chance of getting nothing.

Did you ask yourself which you would prefer?

Most of the respondents preferred the slightly lower probability of obtaining a substantially higher prize—they preferred (*d*) over (*c*).

Surprise, surprise: the choices preferred by the majority—(*a*) over (*b*) and (*d*) over (*c*)—contradict the "independence of irrelevant alternatives," the notorious axiom of von Neumann and Morgenstern, and of Savage... and of anybody else who pretends to think "rationally."

How could this be so? Since this is not immediately obvious, let us rewrite situation (*a*) as lottery (*a′*), noting that "for sure" is equal to 89 percent plus 11 percent. Hence, the decision is between

(a′)   11 percent chance of getting $1 million,
       89 percent chance of getting $1 million,

or

(b′)   10 percent chance of receiving $5 million,
       89 percent chance of getting $1 million,
       1 percent chance of getting nothing.

Because "89 percent chance of getting $1 million" is common to both lotteries, this part of the lottery is irrelevant and, by von Neumann and Morgenstern's fourth axiom, should be ignored. Hence, the question Allais asked boils down to

(a″)   11 percent chance of $1 million,

or

(b″)   10 percent chance of receiving $5 million,
       1 percent chance of getting nothing

Allais's interviewees, and most probably you, the reader, as well, preferred (*a″*) to (*b″*) Note that this choice does not maximize the monetary outcome, but it can be explained by decreasing marginal utility of wealth (i.e., aversion to risk).

So far so good, so now, let's turn to (*c′*) versus (*d′*). We write these as

(c′)   11 percent chance of getting $1 million, and
       89 percent chance of getting nothing,

or

(d′)   10 percent chance of getting $5 million, and
       89 percent chance of getting nothing, and
       1 percent chance of getting nothing.

This time, it is the "89 percent chance of getting nothing" that is common to both lotteries and is irrelevant. It should be ignored. Hence, Allais's second question can be reduced to

(c″) 11 percent chance of getting $1 million,

or

(d″) 10 percent chance of getting $5 million, and
      1 percent chance of getting nothing

Most interviewees preferred (d″) to (c″). Have you, perceptive reader, noticed that (a″) and (b″) are exactly the same as (c″) and (d″)? Nevertheless, very many people prefer (a″) over (b″), but at the same time they prefer (d″) to (c″). What a paradox! By adding "89 percent chance of getting $1 million" to both lotteries, people prefer (a) over (b). But by adding "89 percent chance of getting nothing" to both lotteries, they prefer (d) over (c).

Another way to recognize this paradox is to compute the expected payouts. The expected value of gamble (a) is $1 million, while it is $1.39 million for (b); the expected value of (c) is $110,000, and $500,000 for (d). So, by the theory of expected utility, (b) should be preferred over (a), and (d) should be preferred over (c). In the first gamble, the less risky choice is preferred over a higher expected utility, while in the second gamble, a higher expected utility is preferred over a less risky choice.

One of the people to fall into this trap was Leonard Savage himself. Allais organized a conference in May 1952 in Paris, called "Foundations and Applications of the Theory of Risk Bearing." The "Who's Who" of economic theory—everyone who was anyone—attended. Of course, Friedman (Nobel Prize 1976) and Savage were there, as were Ragnar Frisch (Nobel Prize 1969), Paul Samuelson (Nobel Prize 1970), Kenneth Arrow (Nobel Prize 1972), and many other luminaries.

Over lunch, Allais presented Savage with his questionnaire. The statistician, as knowledgeable about rational decision-making as anybody in the world, considered the situations . . . and promptly chose (a) and (d). Once Allais pointed out his "irrationality," Savage was deeply disturbed—he had violated his own theory! Allais, on the other hand, was jubilant. By challenging the axiom of "independence of irrelevant alternatives," his lunchtime quiz had thrown what he somewhat dismissively called the "American School of rational decision theory," so ably developed by von Neumann, Morgenstern, and Savage, into turmoil.

Allais explained the paradox with the apparent preference for security in the neighborhood of certainty, a profound psychological, if supposedly irrational, reality. As his experiment showed, it superseded the fourth axiom, the independence of irrelevant alternatives.

As there was no room for such lack of rationality in the theory of expected utility, something had to give. "If so many people accept this axiom without giving it much thought," Allais wrote, "it is because they do not realize all the implications, some of which—far from being rational—may turn out to be quite irrational in certain psychological situations."[7] The rational human, behaving according to expected utility theory when faced with risk, simply does not exist.

Several years later, expected utility theory was dealt a further blow with another paradox. This part of the story is linked to a historic event in 1971— the notorious leak of the so-called Pentagon Papers to *The New York Times*, the *Washington Post*, and more than a dozen other news outlets. The Papers, a study commissioned by the U.S. secretary of defense, Robert McNamara, about the American military involvement in Vietnam, noted—among many other deeply embarrassing details—how the war in Vietnam was expanded to Cambodia and Laos without informing the American people, how several U.S. administrations misled Congress, and how several presidents had actually lied to the public. The study was classified as top-secret, and only fifteen copies were made. Two of the copies were given to the RAND Corporation, the think tank in California where Kenneth Arrow had spent summers in the late 1940s. The content raised the profound suspicion of an associate by the name of Daniel Ellsberg (figure 10.1).

Born in 1931 in Chicago, Ellsberg was a highly intelligent boy. His parents were of Jewish descent but had become devout Christian Scientists. His father, an engineer, was unemployed for some time due to the Great Depression but eventually got a job at a renowned engineering firm in Michigan. Family friends remembered the Ellsberg household as cold, with Christian Science front and center. The family prayed every morning, studied a weekly Bible lesson,

FIGURE 10.1: Daniel Ellsberg.

Source: Wikimedia Commons; photo by Christopher Michel.

joined Wednesday night meetings, and attended church every Sunday. The boy excelled at school, but his mother foresaw a career for him as a concert pianist and drove him accordingly. She pushed him to practice six to seven hours a day on weekdays and twelve hours on Saturdays. But when a renowned piano teacher declined to take him on as a student, she was devastated. So great was the maternal pressure that Ellsberg would exclaim later in life, "I've done my time in hell."[8]

But all plans were cut short after a terrible car accident when he was fifteen years old. On their way from Michigan to a family event in Denver, the father, overtired from hours of driving, fell asleep at the steering wheel and crashed into a concrete bridge. He and his son survived, but the mother died instantly, and Ellsberg's younger sister succumbed to her injuries several hours later. Daniel himself was in a coma for thirty-six hours. When he came to, he showed little emotion. It was as if the death of his mother had taken pressure off the boy.

One of the first thoughts that allegedly came to him when he awoke from his coma was that he would no longer have to practice the piano. No wonder he felt the need to turn to psychoanalysis in later life, a decision that would prove to have unintended consequences.

He had no difficulty getting accepted to the college of his choice, Harvard University, and obtained a Pepsi-Cola scholarship to study economics there. After graduating summa cum laude, he spent a year on a Woodrow Wilson Scholarship at the University of Cambridge. In 1954, he joined the Marine Corps and was discharged as a first lieutenant after serving for three years as a platoon leader and company commander, including six months with the Sixth Fleet during the Suez Crisis. Then it was back to Harvard for three years of independent graduate study as a Junior Fellow. In 1962, he earned a PhD in economics at Harvard with a thesis called *Risk, Ambiguity, and Decision*.

While writing his doctoral thesis at Harvard, Ellsberg was employed as a strategic analyst at the RAND Corporation. The choice was not fortuitous because the think tank—at the forefront of the emerging field of decision theory—provided an extremely stimulating intellectual environment. More than thirty Nobel Prize winners have been associated with it to date. But even in the heady environment at RAND, Ellsberg stood out. An officer in the Marines with a Harvard PhD and several papers published in prestigious scientific journals, he was the ideal recruit. Working on classified research that required clearances even higher than top secret, he became acquainted with the processes behind high-level decision-making, such as whether to initiate nuclear war. It was no pretty discovery. What he realized, to his shock, was "that the danger of nuclear war did not arise from the likelihood of surprise attack by either side but from possible escalation in a crisis." This knowledge became "a burden . . . that has shaped my life and work ever since."[9]

Working on an Air Force contract, he became a consultant to the Commander-in-Chief Pacific, then to the Departments of Defense and State, and finally to the White House, specializing in problems of the command and control of nuclear weapons, nuclear war plans, and crisis decision-making. In October 1962, at the onset of the Cuban Missile Crisis, he was called to Washington, and for the next week served on two of the three working groups reporting to the Executive Committee of the National Security Council. A later assignment had him assist on secret plans to escalate the war in Vietnam, although he personally regarded them as wrongheaded and dangerous.

He volunteered to serve in Vietnam, transferring to the State Department in mid-1965. Based at the U.S. embassy in Saigon for two years, Ellsberg's assignment was to evaluate pacification efforts on the front lines. Relying on his Marine training, he accompanied troops to combat in order to see the hopeless war from up close. In the process, he contracted hepatitis, probably on an operation in the rice paddies.

Back at RAND, he worked on the top-secret McNamara study of "U.S. Decision-making in Vietnam, 1945–68," which later came to be known as the Pentagon Papers. This study taught him about the continuous record of governmental deception and unwise decision-making, cloaked by secrecy, under four presidents: Harry S. Truman, Dwight D. Eisenhower, John F. Kennedy, and Lyndon B. Johnson. What was worse, was that he "learned from contacts in the White House that this same process of secret threats of escalation was underway under a fifth president, Richard M. Nixon." His conclusion was that "only a better-informed Congress and public might act to avert indefinite prolongation and further escalation of the war." After meeting conscientious objectors who refused to be drafted to fight in what they saw to be a wrongful war, even at the risk of going to prison, Ellsberg asked himself, "what could I do to help shorten this war, now that I'm prepared to go to prison for it?"

In the fall of 1969, he reached a momentous decision: He photocopied the entire 7,000-page McNamara study and handed it to the chairman of the Senate Foreign Relations Committee, Senator William Fulbright. The senator sat on it without publicizing it, fearing executive reprisal, and in February 1971, the exasperated Ellsberg decided to leak it: first to *The New York Times*, then to the *Washington Post*, and then to seventeen other newspapers. He was sure that he would go to prison for the rest of his life.

The Nixon administration sued to stop the publication of the Pentagon Papers, and the case soon reached the Supreme Court. Citing the First Amendment, the justices ruled 6-3 to uphold the newspapers' right to publish the Papers. Frustrated, the administration tried to cast doubt on the leaker's credibility or otherwise squelch him. Agents acting on the orders of the Nixon administration—they called themselves "the plumbers" because their job was to "plug leaks"—broke into the office of Ellsberg's psychoanalyst and tried to steal the doctor's files on him. Presumably, they hoped either to blackmail him or to smear him so that any future revelations would not be believable. They found nothing because the doctor kept Ellsberg's file under a fake name. There were even plans to physically incapacitate or kill the whistleblower.

Ellsberg faced prosecution on twelve federal felony counts. The indictments carried a maximum sentence of 115 years in prison. This courageous and brilliant young man, with prospects of rising to the highest rung of academia or administration, was willing to risk his career, freedom, and even his life in order to expose the government's deception of its people. Charged with stealing and holding secret documents, he was about to give it all up. But he never wavered. "I felt that as an American citizen, as a responsible citizen, I could no longer cooperate in concealing this information from the American public. I did this clearly at my own jeopardy and I am prepared to answer to all the consequences of this decision," he said in a public statement.[10]

Ellsberg was lucky. During the trial, the judge revealed that representatives of the government had attempted to bribe him (the judge) by offering him the directorship of the Federal Bureau of Investigation (FBI). When, on top of all this, the prosecution claimed that the records of illegal wiretaps had been lost, the judge had had enough. He declared a mistrial and all charges were dismissed "with prejudice," which meant that Ellsberg could not be prosecuted again for his alleged crimes.

Several years before he became notorious for the Pentagon Papers, Ellsberg had made a name for himself in a totally different context. While still a Fellow at Harvard, he published a paper that demonstrated problems with the way in which people handle probabilities. "Risk, Ambiguity, and the Savage Axioms," published in the *Quarterly Journal of Economics*, is considered a landmark in the theory of decision-making.

Following in Allais's footsteps, Ellsberg concocted an experiment. It went like this: suppose that you have an urn that contains ninety balls. Thirty are red, and the other sixty are either blue or green. You do not know how many are green or how many are blue; there may be zero green balls and 60 blue balls, one green and 59 blue, or the other way around, or anything in between. If you guess correctly which ball you draw, you win a prize. But before you draw, you must answer a question: Would you prefer to bet on a red ball or a green ball?[11]

There's no correct answer, of course, and some people would pick green, others would pick red. Then those who answer "red" are asked a second question: Would you prefer to bet that it will be "red or blue," or that it will be "green or blue"? Again, there is no correct answer but many choose "green or blue."

The latter group—be it a majority or just many people—presents a paradox: "If you are in [this group], you are now in trouble with the Savage axioms," said Ellsberg.[12] Let's see why:

The urn contains thirty reds, and sixty other balls that are either blue or green, in some unknown combination.

Does Bertha prefer to bet on a red or on a green? If she opts for red, as some people surely do, she obviously believes that there are fewer than thirty greens in the urn. Hence, by her own belief, the urn must contain more than thirty blues.

Now, does Bertha prefer to bet on "red or blue" or on "green or blue"? If she opts for "green or blue," as some people would, she is irrational. Why? Well, she knows that there are thirty reds, and, as she indicated by her answer to the first question, she believes that there are more than thirty blues. Hence, she should have opted for "red or blue," which, according to her belief, will include more than sixty balls.

Note that when the choice was between red and green, she opted for red. When the choice was between "red or blue" on the one hand and "green or blue" on the other, she opted for the latter. Simply by adding "or blue" to the alternatives made Bertha reverse her decision, thus acting against her own belief!

This is quite amazing. Ramsey had already established the underpinnings of a theory of subjective probabilities. But Ellsberg showed that people behave irrationally, even given their own assessment of probabilities.

The underlying reason for such a situation is that Bertha violated the axiom of the independence of irrelevant alternatives. A rational person, no matter what her belief about the number of green balls, should compare "red or blue" to "green or blue" in the same manner as she compares red to green. The phrase "or blue" is an irrelevant alternative and should not influence her choice, just like adding "89 percent chance of getting $1 million" to both alternatives in Allais's experiment should not influence that decision. It's déjà-vu all over again.

Ellsberg then took his colleagues to task, many more senior than he, either because they participated, and failed, in his experiment, or based on their theoretical work. "There are those who do not violate the axioms, or say they won't, even in these situations. . . . Some violate the axioms cheerfully, even with gusto . . . others sadly but persistently, having looked into their hearts, found conflicts with the axioms and decided . . . to satisfy their preferences and let the axioms satisfy themselves. Still others . . . tend, intuitively, to violate the axioms but feel guilty about it and go back into further analysis. . . . A number of people who are not only sophisticated but reasonable decide that they wish to persist in their choices." The latter group, Ellsberg notes, includes people who felt committed to the axioms and were surprised (even dismayed) to find that they *wished* to violate them.

Ramsey and Savage had surmised that probabilities cannot be considered objective, but Ellsberg proved that even subjective probabilities do not add up. Are people who behave in that manner irrational? That depends on the definition of *rational*. In any case, their behavior is inconsistent with expected utility theory. Ellsberg thought that their choices were motivated by ambiguity aversion: People seem to prefer taking on risk in situations where they know the odds, rather than when the odds are completely ambiguous. When they choose "green or blue," they know that there are sixty balls; had they opted for "red or blue," there could have been anywhere between thirty and ninety.[13] But call it whatever you want, "ambiguity aversion," as did Ellsberg, or "preference for security in the neighborhood of certainty," as did Allais, people's behavior is irrational according to the axioms of game theory. Allais's paradox exhibits a shortcoming of subjective utility, while Ellsberg's paradox demonstrations the inadequacy of subjective probability. Both failings are grounded in the violation of the axiom of the independence of irrelevant alternatives.

CHAPTER 11

GOOD ENOUGH

Traditional economic theory postulates an "economic man," who, in the course of being "economic" is also "rational." This man is assumed to have knowledge of the relevant aspects of his environment . . . [and] also to have a well-organized and stable system of preferences, and a skill in computation that enables him to calculate, for the alternative courses of action that are available to him, which of these will permit him to reach the highest attainable point on his reference scale.[1]

Thus began a paper with the title "A Behavioral Model of Rational Choice," published in 1955 in the *Quarterly Journal of Economics* by one Herbert Simon, a professor at Carnegie-Mellon University of. . . . well, of what field exactly? Variously described as an economist, computer scientist, psychologist, sociologist, and political scientist, the interests of this polymath ran the gamut from public administration, management science, and economics, to artificial intelligence, information processing, and philosophy of science—all of which had in common their connection to the theory of decision-making.

Born in 1916 in Milwaukee, Simon (figure 11.1) was first drawn toward science in high school, though he did not know which kind of science. Neither mathematics, nor physics, nor chemistry, nor biology was quite to his liking. It was human behavior that caught his attention, which he thought was not being studied with scientific precision. "The social sciences, I thought, needed the same kind of rigor and the same mathematical underpinnings that had made the 'hard' sciences so brilliantly successful. I would prepare myself to become a mathematical social scientist," he wrote in an autobiographical piece at the time that he received the Nobel Prize in Economics in 1978.[2] At the University of Chicago, he first intended to study economics, but when he learned that to do so, he would be required to take a course in accounting, he switched to political science.

FIGURE 11.1: Herbert Simon teaching a course.

Courtesy of Carnegie Mellon University

Eventually, he gravitated toward research in public administration. An undergraduate term paper on how organizations make decisions led him to a research assistantship in the field of municipal administration, and then to the directorship of a research group at the University of California at Berkeley. While at Berkeley, Simon completed a dissertation on administrative decision-making at the University of Chicago, taking his doctoral examination by mail.

Jobs were scarce in 1942, but he managed to secure a teaching position in political science back in Chicago, at the Illinois Institute of Technology. That was a stroke of luck, since the Cowles Commission for Research in Economics was then located at the University of Chicago.[3] According to its motto, "Theory and Measurement," the commission aimed to link economic theory to mathematics and statistics. Simon participated in the Cowles Commission's seminars and there, with no fewer than six future Nobel Prize winners on the staff, he obtained a first-class introduction to economics, in particular mathematical economics.

In 1949, Simon moved to the Carnegie Institute of Technology in Pittsburgh, later renamed Carnegie-Mellon University. For half a century, he was instrumental—as a scientist, teacher, and university trustee—in making the school a center of excellence in several fields. He cofounded the Graduate School of Industrial Administration, the School of Computer Science, the Robotics Institute, and the cognitive science group within the psychology department. His reputation as a scholar and founder of several of today's most important scientific fields gained him honors in many disciplines, such as the Turing Award in computer science, the Award for Outstanding Lifetime Contributions to Psychology from the American Psychological Association, the James Madison Award from the American Political Science Association, the von Neumann Theory Prize in operations research, the Dwight Waldo Award from the American Society of Public Administration, the National Medal of Science, and, of course, the Nobel Prize in Economics. Simon died in 2001.

The concepts for which Simon is most famous, and for which he was awarded the Nobel Prize, were *bounded rationality* and *satisficing*. In his quest to understand decision-making, by businesses as well as by individuals, he examined the theories surrounding "economic man" and found them severely lacking: "Recent developments in economics, and particularly in the theory of the business firm, have raised great doubts as to whether this schematic model of economic man provides a suitable foundation on which to erect a theory—whether it be a theory of how firms 'do' behave or how they 'should' rationally behave," he stated in his 1955 paper. Simon sought a new paradigm, continuing: "The task is to replace the global rationality of economic man with a kind of rational behavior that is compatible with the access to information and the computational capacities that are actually possessed by organisms, including man, in the kinds of environments in which such organisms exist."[4]

It was a true paradigm shift. At first, all models of decision-making, from Aristippus and Epicurus onward, postulated that decision-makers maximize something, whether it be pleasure, wealth, or utility. Then, from Daniel Bernoulli onward, the supposition was that people are rational and make rational choices when maximizing the utility of wealth. Next, from John von Neumann and Oskar Morgenstern onward, decision-makers were assumed to adhere to certain axioms. For a long while, game theory held sway and mathematics was the only game in town. Now something new was about to take hold.

For centuries, if not millennia, economics, according to the Merriam-Webster dictionary, was the "description and analysis of production, distribution, and consumption of goods and services," and it was studied solely with the help of narratives, analogies, and anecdotes. With the introduction of the concept of *marginalism* in the late nineteenth century, mathematics first entered the realm of economics—so much so that (as we saw in Chapter 4) Léon Walras thought that he deserved the Nobel Peace Prize for his contribution to the study of economics.

Unfortunately, he was too early. Only in the mid-twentieth century was the time ripe, and in 1968, the Swedish National Bank created the Nobel Prize in Economics. By then, mathematics was the sine qua non of science, and for the first decade or so, the prize seemed off-limits to anybody but a mathematician. After all, mathematical models were the only accepted manner of scientific inquiry, and economists wanted to compete with physicists, chemists, and biologists. If they were to do this without experiments, then at least they had to employ rigorous mathematics.[5] And if the models were based on certain assumptions—namely, markets and competition are perfect, products are homogenous, information is complete, transaction costs are zero, and agents are plentiful and rational—all was OK, so long as these assumptions (even if totally unrealistic) were explicitly spelled out.

But that would not last. Before long, challenges to the status quo loomed. Skeptics, like Maurice Allais and Daniel Ellsberg, took issue with the IIA-axiom (independence of irrelevant alternatives), by showing that normal human beings violate it all the time. With this, they removed one of the indispensable planks on which the modern theory of decision-making was built. And so the tides began to turn. While the traditional mathematical models prescribed how "economic man"—supremely rational, skilled in computing, and all-knowing—*should* act in order to maximize profits or wealth, they were unable to describe how real people actually *do* make these decisions.

Bernoulli had explained the St. Petersburg Paradox with the diminishing marginal utility of wealth. Milton Friedman and Leonard Savage resolved the insurance-gambling paradox with wiggly utility curves, and Harry Markowitz added a wiggle. But legions of further paradoxes, inconsistencies, irregularities, fallacies, and seeming contradictions remained. One author lists dozens of situations where people make seemingly irrational choices:

> [they] display intransitivity; misunderstand statistical independence; mistake random data for patterned data and vice versa; fail to appreciate [the] law of large number effects; fail to recognize statistical dominance; make errors in updating probabilities on the basis of new information; understate the significance of given sample sizes; fail to understand covariation for even the simplest 2×2 contingency tables; make false inferences about causality; ignore relevant information; use irrelevant information (as in sunk cost fallacies); exaggerate the importance of vivid over pallid evidence; exaggerate the importance of fallible predictors; exaggerate the ex ante probability of a random event which has already occurred; display overconfidence in judgment relative to evidence; exaggerate confirming over disconfirming evidence relative to initial beliefs; give answers that are highly sensitive to logically irrelevant changes in questions; do redundant and ambiguous tests to confirm an hypothesis at the expense of decisive tests to disconfirm; make frequent errors in deductive reasoning tasks such as syllogisms; place higher value on an opportunity if an experimenter rigs it to be the "status quo" opportunity; fail to discount the future consistently; fail to adjust repeated choices to accommodate intertemporal connections; and more.[6]

The time had come for mathematicians and economists, until then the prime movers, to make room for new talent. Although the circumstances were not yet ripe for large-scale experiments—Friedman, Savage, Markowitz, Allais, and Ellsberg had based their contributions on introspection and small-scale polls among their friends—it was nevertheless time to turn away from purely theoretical speculation and take humans themselves, not mathematical models of them, as the measure of things.

Rationality, after all, is a matter of definition. So, what about paradoxes and irrationality—the St. Petersburg Paradox, the Allais Paradox, the Ellsberg Paradox, and the insurance-gambling paradox? These are paradoxes only if one postulates the axioms and then decides highhandedly that everyone who does not abide by them is irrational. Simon was the first to take issue with this approach. Maybe the axioms were to be blamed?

Economists define people as rational, so long as they optimize something, like profits, utility, or cost. Mathematicians consider agents as rational, so long as they obey a set of axioms. Psychologists, on the other hand, start by giving human beings the benefit of the doubt: Unless they are obviously demented, whatever humans do is, if not strictly rational, then at least inherently "normal."

Economists accepted that dichotomy by coining two new ideas: Normative economics *pre*scribes, somewhat patronizingly, how "economic man" should behave, in order to maximize or minimize their chance of reaching an objective; meanwhile, positive economics, more cautious, *de*scribes how regular humans, with all their shortcomings and weaknesses, actually *do* behave. Herbert Simon took the latter avenue. He wanted to understand and describe how real humans actually make decisions.

The proverbially rational "economic man"—in more highfalutin jargon: *homo economicus* (see chapter 7, note 2)—is assumed to possess a set of clear-cut preferences, vast knowledge, perfect information, and unlimited computational skills. As the eventual Nobel Prize winner Reinhard Selten would say, "[F]ully rational man is a mythical hero who knows the solutions to all mathematical problems and can immediately perform all computations, regardless of how difficult they are."[7] But alas, preferences are often ambiguous (i.e., utility functions may be multivalued), access to information is imperfect, and computational capacities are limited. Simon soon came to a profound realization: human beings are unable to do what is expected of "economic man." This is why they get entangled in what seem like paradoxes.

Take, for instance, the price of a call option. Merchants have been trading options on all kinds of commodities for centuries. Throughout history, buyers and sellers relied on their gut feelings to determine the price of options. Then, in 1973, Fisher Black, Myron Scholes, and Robert Merton showed that the correct price of a call option, the price that rational traders should be paying for it, is given by the following equation:[8]

$$C(S, t) = N(d_1)S - N(d_2)Ke^{-r(T-t)},$$

where

$$d_1 = \frac{Ln\left(\frac{S}{K}\right) + \left(r - \frac{\sigma^2}{2}\right)(T - 1)}{\sigma\sqrt{T - t}}.$$

$$d_2 = d_1 - \sigma\sqrt{T - t}$$

I won't even bother explaining the symbols and variables, except to mention that σ stands for the volatility of the underlying share price, which can be estimated only with fairly sophisticated tools based on historical data. Do economists truly believe that rational traders buy and sell options at prices that correspond to such a complex equation? True, the pricing of options may be an extreme example, but even simple choices, like whether to take an umbrella when leaving the house, require the collection and storage of data, an assessment of probabilities, the computation of costs and benefits, the processing of these results, and finally—often with lightning speed—the making of a decision.

It seems patently absurd to suppose that anybody would actually go through such a process. Hence, Simon concluded that "actual human rationality-striving can at best be an extremely crude and simplified approximation to the kind of global rationality that is implied, for example, by game-theoretic models."[9] So, regrettably, the behavior of human decision-makers simply does not conform to the models developed by the likes of von Neumann and Morgenstern and accepted by economists smitten with the beauty of their mathematics. To Simon, the models were acceptable as instructions—how to maximize value, wealth, or utility—but as descriptions of how humans go about deciding things, they had to be discarded.

In contrast to Allais and Ellsberg, however, Simon's scorn was not motivated by the refutation of any of the axioms; he rejected the supposedly rational mathematical models because of the burdens they place on the decision-maker, as well as the limitations of human capabilities. It was not the alleged irrationality of human beings that caused him to discard the mathematical model, but rather his understanding that the need to gather data and perform complex computations exceeded the skills of human beings. Human decision-makers were not irrational, he claimed, but cognitive limitations rendered them *boundedly rational*. Within their bounds, they were rational.

What did Simon propose instead? To provide a more realistic description of decision-making by humans, he said, one must forsake the idealized versions of economic man and inspect the human mind itself. To do so, answers needed to be sought in psychology, not mathematics.[10] Hence, the rationality of "economic man" was to be replaced with "a kind of rational behavior that is compatible with the access to information and the computational capacities that are actually possessed by organisms, including man, in the kinds of environments in which such organisms exist," Simon wrote in his groundbreaking 1955 paper.[11]

The availability of too many choices, the difficulties in analyzing them, and the lack of time to process everything cause people not to seek the very best alternative, he proposed. Instead, when faced with the task of finding the elusive optimal choice, human beings take a shortcut: Making efficient use of their abilities and time, they identify an acceptable option and go with it. They decide on a level that is good enough to satisfy their needs, and then they choose the first option that meets or surpasses it. Hence, they do not optimize, but rather seek a solution that suffices, that satisfies their aspirations—that is, they *satisfice*.

But there's more. Not only do decision-makers avoid the onerous search for the true optimum, but to find and identify the satisficing, albeit possibly suboptimal, choice, they often employ *heuristics*, which is a fancy word for "rules of thumb." For example, when determining the growth by two small percentages, it may suffice to add them. To illustrate, a growth first by 10 percent and then by 40 percent may be approximated by a growth by 50 percent, although the correct growth is $1.1 \times 1.4 = 1.54$ (i.e., 54 percent). Because most people find it easier to add than to multiply, the slight error in the result is offset by the lower cost of deliberation. For boundedly rational individuals, such shortcuts often provide a satisfactory solution without the burden of costly, time-consuming calculations that they may not even be able to perform anyway.

So, after criticizing the theory of expected utility and game theory, Simon suggested the theory of bounded rationality. Analyses based upon optimization should be replaced by models based upon satisficing and heuristics.

It should be noted that bounded rationality is not at all the same as irrationality. When considering all the costs, the use of heuristics is quite rational because rules of thumb allow the identification of satisfactory solutions without the need to embark on onerous, often impossible, mental tasks. Hence, by including the costs of data collecting, information processing, and deliberation of all facts in the total cost of the search, it is quite *rational*, in the traditional sense of the word, to use heuristics to satisfice in a cost-effective manner. It is simply a tradeoff between the search for the optimum and the cost of a reasonable deliberation.

People develop heuristics through learning, experience, interaction with the environment, and feedback. Often, heuristics come under the guise of *common sense* or *educated guesses*. Unfortunately, while rules of thumb reduce mental effort, it is unavoidable that they introduce errors and distortions. We shall see in Chapter 12 that such mistakes, commonly called *biases*, are exhibited by most people, usually in a systematic manner—much like what occurs when the product of two positive numbers is replaced by a sum, a rule of thumb that *always* underestimates the result.

What kind of heuristics do human beings utilize? One need only look at the seemingly irrational procedures that people employ to make choices. To mention some of them: People use the information at hand, be it pertinent or immaterial, round low probabilities down to zero and high probabilities up to 100 percent, believe evidence if it confirms their prejudices, exaggerate the importance of vivid over bland evidence, ignore new information, assign high probability to random events that are fresh in their memory, rely with overconfidence on their own judgment, discern apparent patterns in random data, and, alas, fail to ignore irrelevant alternatives.

CHAPTER 12

SUNK COSTS, THE GAMBLER'S FALLACY, AND OTHER ERRORS

As we saw in Chapter 11, people routinely use rules of thumb to evaluate situations and make decisions. Known as *heuristics*, they serve as shortcuts, especially when decisions must be made in uncertain situations. Herbert Simon first proposed heuristics in the mid-1950s, when he proposed bounded rationality and satisficing as an alternative to the maximization of expected utility. Others, from Maurice Allais onward, had already contested expected utility theory (or, rather, the expected utility *hypothesis*). But it was a paper written two decades later, by the Israeli psychologists Amos Tversky and Daniel Kahneman, that had the greatest impact. Published in 1974 in *Science*, the premier scholarly journal in the United States, it hit not only the economics profession like a bombshell, but the world of science in general. Taking issue with the received wisdom of rational, economic man, it put paid to the very idea of the expected utility theory.

"People rely on a limited number of heuristic principles which reduce the complex tasks of assessing probabilities and predicting values, to simpler judgmental operations," they stated, adding that, unfortunately, the convenience and the savings in time and effort that heuristics afford come with a price: The results will be inexact. "In general, these heuristics are quite useful but sometimes they lead to severe and systematic errors."[1] But there was a glimmer of hope. The word *systematic* is key in the preceding quote. What the authors found was that the errors and biases were not random—they usually went in the same direction and could be rigorously tracked and studied.

SUNK COSTS, THE GAMBLER'S FALLACY, AND OTHER ERRORS   189

Amos Tversky and Daniel Kahneman were considered stars among the scientists of their day. A recent best-seller about their lives, their collaboration, their friendship, and their eventual falling-out attests to the significance of their work.[2]

Born in Palestine in 1934, Kahneman (figure 12.1) spent his childhood years in Paris, where his father was a chemist for the cosmetics company L'Oréal. When Jews were rounded up and taken to the infamous Drancy prison, from which they were transported to concentration camps, Kahneman's father was among them. But he was lucky. His employer, later exposed as a Nazi sympathizer, got Kahneman's father released by deeming his work indispensable for L'Oréal. Shortly thereafter, the family fled Paris for southern France. They moved around, constantly in fear of being discovered by Nazis, Vichy collaborators, or private bounty hunters seeking out Jews. Eventually, they found refuge in a chicken coop in a village outside Limoges, where the father died of diabetes. Daniel and the rest of his family survived the war and moved to Palestine, several months before the creation of the state of Israel.

After high school, Kahneman entered Hebrew University in Jerusalem to study psychology. Upon graduation, he performed his compulsory military service with the Israeli air force. As a psychologist, he was asked to devise tests to evaluate candidates for officers' courses. Following his service, he went to the University of California at Berkeley to write his doctoral thesis and obtained his degree in 1961.

Amos Tversky (figure 12.2), Kahneman's junior by three years, was born in Haifa, part of mandatory Palestine, in 1937. His father, a veterinarian, and his mother, a politically active social worker, belonged to the founding generation of the Jewish state. As a representative of the Israeli Workers' Party, the mother became a lawmaker in the Knesset, the Israeli parliament.

When Tversky was drafted into the army, he volunteered for the notoriously tough paratroopers, where he became known for his bravery. Once, in a training exercise, a soldier pulled the trigger of an explosive device but then froze in shock, before he could run for cover. Disregarding his own safety, Tversky sprinted toward him and pulled him away. The feat earned him a medal for bravery, not to mention shrapnel that remained in his body for the rest of his life. He completed his compulsory military service as a captain, and following

FIGURE 12.1: Daniel Kahneman.

Source: Wikimedia Commons, © Renzo Fedri

FIGURE 12.2: **Amos Tversky.**

© Ed Souza/Stanford News Service

that, he studied psychology and mathematics at Hebrew University. In 1964, Tversky obtained his PhD in psychology from the University of Michigan. He remained a reserve officer of the Israeli Defense Forces, however, and years later, he had to take off from his research and teaching to fight in the Six-Day War in 1967, and then in the Yom Kippur War in 1973.

The two psychology professors met at Hebrew University when Kahneman invited his younger colleague to give a talk at his graduate seminar. Their collaboration began with a disagreement about the content of the lecture. It was nothing less than amazing that two men of such different character—Tversky, brash and brilliant, Kahneman, reticent and full of self-doubt—could work together so prolifically and (mostly) harmoniously. Their partnership become something of a legend. For years, they seemed inseparable, spending hours together behind closed doors, day after day, investigating the ways of the human mind. They produced groundbreaking research and seemed to have the best of times doing it. Only their laughter could be heard outside their office.

After many years of collaboration, however, a cloud gathered. Tversky had obtained a professorship at Stanford, while Kahneman got a position at the well-known, but much less prestigious, University of British Columbia. In general, the garrulous Tversky outshone his restrained friend. Recipient of a MacArthur Fellowship, fellow of the American Academy of Sciences, member of the National Academy of Sciences—Tversky got most of the honors, which became, for a while, a source of friction. By the time Tversky died of cancer in 1996, at only fifty-nine years old, they had made up again. Six years later, in 2002, Kahneman, then at Princeton University, was awarded the Nobel Prize in Economics "for having integrated insights from psychological research into economic science, especially concerning human judgment and decision-making under uncertainty," work that he and Tversky had done together.[3] But he was magnanimous in his triumph: The first sentence of his Nobel Prize Lecture was, "The work cited by the Nobel committee was done jointly with the late Amos Tversky (1937–1996) during a long and unusually close collaboration."[4]

As science Nobel prizes go, the theories proposed by Kahneman and Tversky may be among the easiest—indeed, possibly the only ones—that laypeople can understand. With the possible exception of the work by Simon, the award-winning achievements in chemistry, physics, and economics are generally chock-full of mathematical equations. Kahneman and Tversky's work, in contrast, does not go beyond some probability theory and statistics.

In their groundbreaking 1974 paper, "Judgment Under Uncertainty: Heuristics and Biases," they proposed and analyzed three heuristics that we all use to make decisions—generally without even realizing it. They are the *representativeness heuristic*, the *availability heuristic*, and the *anchoring heuristic*.

The most famous example of the representativeness heuristic, and how it can lead one astray, is the case of "Linda, the bank teller." It is based on an experiment in which subjects were given some information and then asked to make an evaluation:

Linda is thirty-one years old, single, outspoken, and very bright. She majored in philosophy. As a student, she was deeply concerned with issues of discrimination and social justice, and also participated in antinuclear demonstrations. Which is more probable?

1. Linda is a bank teller.
2. Linda is a bank teller and is active in the feminist movement.

When Tversky suggested the experiment, Kahneman was skeptical. Obviously, the correct answer was "Linda is a bank teller." This was not an opinion, but a mathematical fact. After all, "bank teller" is a broad category that includes feminists. Hence, whatever the probability that Linda is a bank teller, the probability that Linda is both a bank teller and a feminist is necessarily smaller.

Against his better judgment, Kahneman ran the experiment. To his immense surprise, most subjects gave response number 2. In their opinion, Linda was more probably a bank teller *and* a feminist, rather than just a bank teller.

Kahneman and Tversky concluded that people make a mental representation of Linda according to the information given. The average bank teller is nothing like Linda, if only because a large fraction of bank tellers are men (at least, they were at the time of this experiment). However, of female bank tellers who are feminists, Linda is a fairly good representative.

The two psychologists had hit on a profound truth. People tend to go with the representativeness of the description, not with the logic. If a description of a person is, say, very representative of a four-star general of the U.S. armed forces (e.g., physically fit, stern, highly accomplished, disciplinarian, etc.), people may think that this person is a general, rather than a high-school gym teacher, even though this is highly unlikely to be true because only two dozen such flag officers are active in the American military at any time.

In his Nobel Lecture, Kahneman gave a further example. When evaluating a set of objects, its basic representation includes average or typical values, but not more complex statistics like sums.[5] Thus, the representativeness heuristic replaces a set's sum by its average. In an ingenious experiment, subjects were presented with a set of eight plates and asked to judge the set's value.[6] Then they were presented with another set that contained not only the same eight plates, but four more, two of which were cracked.[7] Surprisingly, people by and large would judge the set that includes the cracked plates to be worth less than the former.

Again, this is quite wrong mathematically. Even though the second set contains two cracked (and therefore worthless) plates, it does contain ten perfectly good ones and must therefore be worth more than the eight in the first instance. What happens, however, is that people unwittingly average out. Let's say that they judge the set of eight plates to be worth $8 (i.e., each plate at one dollar). Then the ten good plates in the second set should be worth $10, while the two cracked plates, worth nothing, should simply be ignored. But the *average* value of each of the twelve plates in the second set is only about 83 cents ($10 divided by 12). So, by considering the average as representative of the set's value rather than the sum, which is a more complex characteristic, the second set is incorrectly judged to be worth less than the first.

Both Linda and the cracked plates are violations of the vexatious axiom of "independence of irrelevant alternatives" all over again. If the question had been "Linda is a woman; is she more likely to be a bank teller or a bank teller and a feminist?" most people would opt for bank teller, without the qualifier. By adding nondescript, irrelevant information—outspoken, bright, antinuclear, etc.—many change their minds. The same holds for the cracked plates scenario. Instead of simply ignoring the cracked plates, the subjects thought that their addition to the set of plates reduced their valuations.[8]

Next, we turn to the availability heuristic. It does what it says it does: utilize the information that's available, relevance be damned. Well, not quite—but to evaluate situations, people usually consider simply what's at hand. Frequencies and probabilities are judged by the ease with which pertinent examples come to mind, be it because they are vivid, well publicized, or recent. This might be a coping mechanism that has evolved throughout human history, though Kahneman and Tversky consider a shorter time span. "Lifelong experience has taught us," they surmised, "that, in general, instances of large classes are recalled better and faster than instances of less frequent classes; that likely occurrences are easier to imagine than unlikely ones; and that the associative connections between events are strengthened when the events frequently co-occur." Hence, the availability heuristic offers procedures to estimate "the numerosity of a class, the likelihood of an event, or the frequency of co-occurrences, by the ease with which the relevant mental operations of retrieval, construction, or association can be performed."[9]

Unfortunately, such shortcuts lead to errors. It's rather like searching where the light is.[10] For example, when asked whether there are more English words that begin with a *k* than have a *k* in the third position, people judge the former to be more likely because words like *knee* or *key* come to mind more easily than words like *acknowledge* or *like*. In fact, however, there are three times as many English words that have *k* in the third position.

Another, much more serious—indeed, fraudulent—misjudgment was the hoax by a certain Dr. Andrew Wakefield, a medical quack in England, whose license to practice has since been revoked. He claimed that the vaccination of children leads to autism. However, it so happens that vaccinations are administered to children when they are about two years old, which is just about the age at which symptoms of autism first appear. This is a coincidence, and as was exhaustively proved later, the correlation is illusory. Unfortunately, naive parents confuse coincidence with evidence and are sufficiently convinced to leave their children unvaccinated.

Next, let me cite the widespread fear of flying as an example of the availability heuristic. There are far, far fewer victims of air accidents than there are of road accidents. But the attention given by the media to the former is far, far greater—hence the pervasive aviophobia. And, while we are speaking of traffic accidents, when people were asked in a study which causes more fatalities, vehicle accidents or lung cancer, 57 percent opted for the former, even though death from lung cancer is three times as frequent. But car accidents are often reported—even if less emphatically than air accidents—while deaths from lung cancer hit the news only rarely. In the words of the two psychologists, "the ease with which disasters can be imagined need not reflect their actual likelihood."[11]

The third rule of thumb that Kahneman and Tversky discussed in their paper is the anchoring heuristic. Have you ever noticed that many high-class restaurants offer some really expensive items on their wine list that nobody would even think of ordering? Well, neither would the sommelier—the bottle is listed for a different reason. Once a patron sees the Chateau Lafite 1865 for $5,000 on top of the list, the 2008 Domaine Leflaive Puligny-Montrachet Les Folatières 1er Cru farther down, at $250, seems like a steal. This is an example of the anchoring heuristic. Faced with an arcane question, decision-makers latch onto a number, however irrelevant it may be, which then becomes the new normal.

Kahneman and Tversky put this heuristic on a sound footing. In another one of their experiments, they asked their subjects what the percentage was of African states represented in the United Nations. At the start of the session, the experimenter spun a wheel of fortune that came to rest either on 10 or on 65, and then the participants were asked the question. Lo and behold: The average answer of the group who had been primed with a 10 thought that African nations made up 25 percent of the UN membership, while subjects primed with 65 put the number at 45 percent. (The correct answer today would be 28 percent, 54 out of 194 members.) Obviously, the subjects anchored their initial estimates at the wheel-of-fortune number and then adjusted their answers up or down.

In another test, groups of high school students were asked to estimate, within 5 seconds, the product $1 \times 2 \times 3 \times 4 \times 5 \times 6 \times 7 \times 8$ and $8 \times 7 \times 6 \times 5 \times 4 \times 3 \times 2 \times 1$. The answers differed markedly: The median estimate was 512 for the ascending sequence and 2,250 for the descending sequence. (The correct answer—for both, of course, thanks to the commutative property—is 40,320.) Apparently, the students anchored themselves by performing the first few multiplications, and then adjusted the result. Because the first few multiplications in the ascending sequence result in a much lower number than in the descending sequence, the first group gave a much lower estimate.

The representativeness heuristic is often employed when people are asked to judge the probability that an object or event belongs to a certain class, the availability heuristic is used to assess the frequency of a class or the plausibility of a development, and the anchoring heuristic is used when making numerical predictions. Unfortunately, these rules of thumb lead to systematic and predictable errors, called *biases*, which may lead people astray.

A very common bias is the so-called *sunk cost fallacy*. A person who has invested time, money, and effort in an endeavor may be loath to abandon it, even if it turns out that it is no longer profitable, or a better alternative has become available. He or she tends to remain with the original choice because of the costs already expended. This decision is incorrect, though, because whatever has already been expended is a *sunk cost;* it won't come back even if you abandon the project. Hence, sunk costs should never influence one's decision. One must only look forward, not backward.

Failure to recognize *regression to the mean* is another source of error. Kahneman described an encounter with the commander of a fighter pilot school in Israel. The commander grumbled that whenever a pilot is commended for an especially good execution of a maneuver, he invariably performs worse the next time, and vice versa. Kahneman enlightened him about a universal fact—namely, whenever something is exceptionally good, large, heavy, or fast, the next instance tends to be worse, smaller, lighter, or slower, etc. It is the very definition of a mean that outliers tend to be followed by less extreme values.

Regression to the mean is observed whenever a variable depends on chance but has a well-defined average. For example, the children of fathers with exceptionally low IQs tend to have IQs higher than their fathers. To show this, one can draw a bell curve, with an IQ equal to 100 at the center. Now draw a vertical line at the father's IQ of 85. Obviously, the area under the bell curve to the right of 85 is much greater than to the left of 85, which means that the children most probably will have IQs greater than that of the father. Their IQs regress toward the mean of 100.[12]

One question that needs to be addressed is: What is the variable that one needs to look at? Ever since the early eighteenth century, when Daniel Bernoulli was wrestling with the St. Petersburg Paradox, a person's utility was expressed in terms of wealth. Beginning at zero wealth, the utility curve continually rises, albeit at a decreasing rate. That was the subject matter of all the chapters of this book so far. Kahneman and Tversky took issue with wealth being the relevant variable. After all, who is happier: he who had $4 million yesterday and lost $1 million, or she who had $1 million yesterday and gained $100,000? It is not absolute wealth that determines one's utility for money, they claimed, but gains and losses. The point of reference should be set at current wealth, and decisions about money matters should be analyzed in terms of changes from that customary point.[13]

Another well-known reason for biased decisions is framing. The manner in which a question is framed may elicit contradictory answers. Consider the following experiment on which Kahneman and Tversky reported in *Science* in 1981:

Imagine that the U.S. is preparing for the outbreak of an unusual Asian disease, which is expected to kill 600 people. Two alternative programs to combat the disease have been proposed. Assume that the exact scientific estimate of the consequences of the programs are as follows:

If Program A is adopted, 200 people will be saved.
If Program B is adopted, there is 1/3 probability that 600 people will be saved, and 2/3 probability that no people will be saved.

When asked which program they prefer, three-quarters of the subjects chose Program A. Then, another group was asked the same question but it was formulated differently:

If Program C is adopted, 400 people will die.
If Program D is adopted, there is 1/3 probability that nobody will die, and 2/3 probability that 600 people will die.

This time, three-quarters chose Program D. Obviously, Programs A and C are identical, and so are Programs B and D. But "200 will be saved" strikes a positive note, while "400 will die" sounds very undesirable.

This and many similar experiments led Kahneman and Tversky to two conclusions. First, subjects can be pushed to give a desirable answer through judicious formulation of the question. For example, credit card companies insist that retail shops that want to entice customers to pay cash frame the difference between the gross and the net not as a credit card surcharge, but as a cash discount. Surcharges are seen as outlays, discounts as opportunity costs.

Second, if the alternatives are formulated in terms of gains, as in the first instance, the majority of subjects opt for the certain outcome (i.e., they are risk averse). If the alternatives are framed as losses, as in the second instance, they opt for the uncertain alternative (i.e., they are risk taking). This becomes even more noticeable when subjects are asked, "Do you prefer a sure gain of $10,000 or a 50 percent chance of getting $20,000?" and "Do you prefer a sure loss of $10,000 or a 50 percent chance of losing $20,000?" In the first case, most people are risk averse and opt for the sure gain; in the second case, most are risk seeking and opt for the 50 percent chance of a loss. In short, individuals are more willing to take risks to avoid a loss than to secure a gain.

Next, the *gambler's fallacy* is the misperception about "what has to happen next" in random events, like spinning a roulette wheel or tossing coins. In the coin-tossing version, it goes back to the St. Petersburg Paradox. Many gamblers falsely believe that if a coin fell heads, say, five times in a row, tails is due on the next toss in order to bring the total record back into kilter. In fact, this is totally incorrect. Tosses are completely independent, and coins do not "remember" their history. Hence, the chance of a coin falling on tails on future tosses is, and always remains, 50 percent. So, what about the regression to the mean? Should the coin not fall on tails, to even out the string of random tosses? No, that argument has no bearing on the next coin toss or roulette spin at all. Yes, the average over many coin tosses is 50 percent heads and 50 percent tails, but this is a statement about a group of tosses and spins. The regression to the mean holds true only over many repetitions, while the gambler's fallacy is a false statement about the single, next event.[14]

The coin-tossing scenario can illustrate another kind of bias. When asked which of the two strings of random coin tosses, TTTTTT or THHTHT, is more likely, most people opt for the latter because it looks more random. In fact, basic probability theory says that both strings are equally likely. Nevertheless, many people expect small samples to echo the characteristics of a much longer run. To them, THHTHT just seems more representative of a random sequence.

These and many other biases led Kahneman and Tversky in 1979 to formulate an entirely new model of decision-making. It took no little amount of audacity to suggest that traditional decision theory, from Bernoulli's utility of wealth up to von Neumann and Morgenstern's axioms, needed to be replaced. Even more disconcerting to many traditionalists was the fact that these two authors weren't even economists, though they were well versed in probability, statistics, and mathematical modeling. Hence, for two Israeli psychologists to question centuries of received wisdom and to take on the leading lights in economics must have seemed like the height of chutzpah. Nevertheless, "Prospect Theory: An Analysis of Decision Under Risk," published in one of the leading academic journals of the profession, *Econometrica*, became hugely influential. It became the new theory, not about how people *should* make decisions under uncertainty, but how they *do* make them. According to Google Scholar, it has been cited nearly 50,000 times.

The tone is set in the paper's abstract, where the authors announced that "this paper presents a critique of expected utility theory as a descriptive model of decision-making under risk." Even after that bombshell proclamation, they did not let up. "[U]tility theory, as it is commonly interpreted and applied, is not an adequate descriptive model and we propose an alternative account of choice under risk." They named this alternative model *prospect theory*. The word *prospect* refers to uncertain situations, like investment proposals, lotteries, or insurance contracts.[15]

According to prospect theory, decisions under risk proceed in two stages. In the first, called the *editing process*, decision-makers translate the problem at hand into something more tractable by the human mind. To do so, heuristics are employed, with all the bells and whistles and biases. In the second stage, the *evaluation process*, they appraise the alternatives and choose the most favorable option. To do this, decision-makers use the utility curve and a weighting function that is roughly based on probabilities.

The editing process consists of several operations: (1) coding, (2) combination, (3) segregation, (4) cancellation, (5) simplification, and (6) detection of dominance. All of these are conducted more or less subconsciously.

(1) The first break with Bernoulli's worldview comes in the coding operation. The Swiss mathematician, as well as those who came after him, had formulated their models in terms of final wealth. But recall who is happier: he who had $4 million yesterday and lost $1 million, or she who had $1 million yesterday and gained $100,000? The answer to this question, elicited in numerous experiments by Kahneman and Tversky, was that the final wealth position is irrelevant. Rather, decision-makers consider gains and losses relative to some neutral point as the appropriate variable.

(2) Decision-makers simplify prospects by combining the probabilities of events with identical outcomes. For example, a 30 percent chance of winning $100 and a 20 percent chance of winning $100 will be combined to "50 percent chance of winning $100."

SUNK COSTS, THE GAMBLER'S FALLACY, AND OTHER ERRORS   201

(3) The riskless components of a prospect are segregated from the risky part. A gamble with a 50–50 chance of getting either $180 or $300, for example, would be segregated into "a certain gain of $180, and a 50–50 chance of gaining another $120 or nothing." Similarly, a 50–50 chance of losing $60 or $280 is translated into "a sure loss of $60 and a 50–50 chance of losing nothing more or another $220."

(4) If there are two prospects with a common element,

(a) 20% of + 230, 30% of + 260, and 50% of −70,
(b) 20% of + 230, 30% of + 360, and a 50% of − 135,

the decision-maker will segregate the common element, 20 percent of +230, and remain with a choice between the following two:

(a') 30% of + 260, and 50% of −70,
(b') 30% of + 360, and a 50% of − 135.

(5) *Simplification* refers to the rounding of probabilities and outcomes. An event that has a 49 percent chance of paying $101 will often be rounded to a 50 percent chance of getting $100. And extremely low or extremely high probabilities will be rounded to zero or to 100 percent, respectively. In other words, such events will be discarded or treated as certain events.

Finally, (6) alternatives that are dominated (i.e., they have both smaller outcomes and lower probabilities than other alternatives) are rejected out of hand.

Once the editing process has been completed, with all its faults, biases, and shortcomings, the decision-maker proceeds to the second phase, the evaluation process.[16] Mathematically, it looks very much like the expected utility model... except that it's nothing like the expected utility model.

According to raw economics, the way to decide among alternatives is to maximize expected wealth. Final wealth positions in dollars are weighted by the probabilities of their occurrence, and the alternative with the highest expected wealth is chosen. As a prescription for maximization of economic wealth, this is still the correct method. But people, being people, behave differently. With insight into human nature, Bernoulli replaced "wealth in ducats" with "utility of wealth in ducats." According to that model, the alternative with the highest "expected utility" is chosen. And, of course, the curve that describes utility was assumed to rise constantly, albeit at a decreasing rate. That was the model that was accepted for the following two centuries, until Friedman, Savage, and Markowitz added some wiggles to the utility curve.

Then came Kahneman and Tversky, who tweaked common utility in several ways. Instead of the customary utility function, they defined a "value function" that has three essential features. First, human beings consider changes in wealth when making decisions. Hence, the value function is defined not in terms of final wealth, but in terms of gains and losses. To back up that claim, the authors not only cite their own experiments with students from all over the world, but also refer back to Ernst Heinrich Weber and Gustav Theodor Fechner (see Chapter 3), whose experiments with weight, loudness, temperature, and brightness showed that, in general, stimuli are perceived in relation to a reference point.[17] Hence, Kahneman and Tversky's value function reflects the subjective value of deviation from current wealth.

Second, as we saw at length when I discussed diminishing marginal utility in the early chapters of this book, people experience the difference in value between a gain of 100 and a gain of 200 as greater than the difference between a gain of 1,100 and a gain of 1,200. Similarly, the difference between a loss of 100 and a loss of 200 appears greater to decision-makers than the difference between a loss of 1,100 and a loss of 1,200. The general conclusion is that the value curve—like its predecessor, the utility curve—is concave (i.e., risk averse) for gains, and convex (i.e., risk loving) for losses.

Third, an important characteristic of attitudes to changes in wealth is that "losses loom larger than gains. The aggravation that one experiences in losing a sum of money appears to be greater than the pleasure associated with gaining the same amount."[18] In general, gains that are certain are favored over probabilistic gains, and probabilistic losses are preferred to definite losses. For example, Kahneman and Tversky maintain that most people reject bets that give equal odds of gaining or losing $50. This implies that the value function is generally steeper for losses than for gains (figure 12.3).

So far, it is all pretty humdrum. Where Kahneman and Tversky really parted ways with their predecessors was in the way that they determined that probabilities must be treated. We know that due to the vagaries of the rules of thumb that people use to estimate probabilities, their stated probabilities are subject to biases and errors. But the two psychologists found something more: Their experiments showed that people transform stated probabilities

FIGURE 12.3: The Kahneman and Tversky value function.

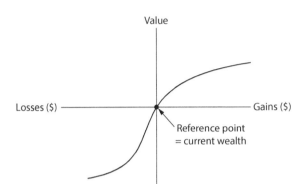

into what they called decision weights: "Decision weights measure the impact of events on the desirability of prospects, and not merely the perceived likelihood of these events."[19]

The transformations occur according to certain rules. Although decision weights act much as probability does in the expected utility framework, Kahneman and Tversky identified several profound differences between the two measures. Based on their experiments, they recognized that people treat decision weights as (1) subcertain, (2) subadditive, and (3) subproportional. (I will explain these notions next.) In addition, (4) small probabilities are overweighted, except for extremely low probabilities. Finally, (5) events whose occurrence is extremely likely or extremely unlikely never make it to the evaluation phase; they are already handled in the editing phase.

(1) *Subcertainty* refers to the fact that the decision weights that people accord to complementary events do not add up to 100 percent. Take the Allais Paradox, for example, with probabilities of 89 percent and 11 percent. We denote probability by p, the decision weight accorded to it as $\pi(p)$. While probabilities 89 percent and 11 percent add to 1.0, the decision weights $\pi(89 \text{ percent})$ and $\pi(11 \text{ percent})$ in the framework of the Allais Paradox do not. Kahneman and Tversky's subcertainty provides an explanation of how the Allais Paradox may come about.[20]

(2) *Subadditivity* was discovered in experiments in which subjects were asked if they preferred, say, a 0.1 percent chance of receiving $6,000 or a 0.2 percent chance of receiving $3,000. Most people preferred the former, even though the expected payouts are identical. This implies that people value a chance of 0.2 percent as less than twice 0.1 percent, despite the reality (i.e., decision weights are subadditive).[21]

(3) *Subproportionality* is subtler.[22] Suffice it here to say that subproportionality imposes considerable constraints on the shape of the decision weight $\pi(p)$: its logarithm must be a convex function of the logarithm of the probability.

(4) Small probabilities are generally overweighted.[23] This is borne out, for example, by the fact that most people prefer a 0.1 percent chance of obtaining $5,000 to a certain receipt of $5.[24] Hence, for small values of p, $\pi(p) > p$.

(5) Near the end points, at probabilities close to zero and close to 100 percent, decision weights are not well defined. Events with such extreme probabilities are already dealt with at the outset, in the editing phase: Events with probability of, say, 0.01 percent are considered "never to happen" and are ignored, in effect setting their decision weights equal to zero. Similarly, events with probability of, say, 99.99 percent are considered "sure things," in effect setting their decision weights to 100 percent.

Based on these findings, the curve that describes decision weights must look something like figure 12.4.

In this chapter, we discussed the modern theory of judgment and decision-making. Classical economics posited that *homo economicus*, the assumed player in the mathematical models, is rational, maximizes expected utility, is free of emotion, and makes no mistakes when gathering and processing information. But *homo sapiens*, the real-life player in today's environment, is not at all like that. Herbert Simon (see Chapter 11) was among the first to recognize the incongruity of the classical assumptions with what takes place in the real world. He asserted that limited abilities and imperfect skills make the mathematical models—as useful as they are for normative economics—all but useless for the description of what people actually do. *Bounded rationality* and *satisficing* were the new catchwords.

FIGURE 12.4: Kahneman and Tversky decision weights.

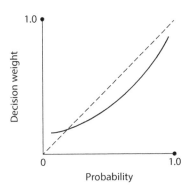

But there are more limitations than bounded rationality. It is a fact of life that people are subject to cognitive biases. This is where Daniel Kahneman and Amos Tversky came in. Judgment is about estimating magnitudes and probabilities. These two psychologists found that human beings use heuristics (i.e., rules of thumb) to obtain, elicit, or guess data. Prospect theory is about how the information thus acquired is utilized to make decisions. Based on experiments, Kahneman and Tversky identified certain idiosyncratic properties of value functions and decision weights that people employ to make decisions under risk.[25]

So the previous thinking was that the models are perfect, and even though people make mistakes occasionally, these mistakes wash out on average in the market. Kahneman and Tversky showed that this was not so: people make systematic mistakes. . . .in the same direction. Hence, they do not wash out. The market is systematically "wrong."

It was a true paradigm change.

CHAPTER 13

ERRONEOUS, IRRATIONAL, OR PLAIN DUMB?

When Daniel Bernoulli investigated the St. Petersburg Paradox in the early eighteenth century, he used mathematics to do it, albeit with remarkable insight into the human mind. In the later eighteenth century, Adam Smith, in his *Wealth of Nations* published in 1776, took a decidedly more guarded view of mathematics. "I have no great faith in political arithmetick," he asserted, underpinning his theories with examples, narratives, anecdotes, and literary references instead.[1] In fact, in *The Theory of Moral Sentiments*, he proposed psychological explanations of individual behavior, and even expressed concerns about fairness and justice.

The utilitarian Jeremy Bentham wrote extensively on the psychological underpinnings of utility. In the late nineteenth century and continuing until about the 1970s, it was déja vu all over again (to quote Yogi Berra): Once more, mathematics, the queen of the sciences, reigned. First, the marginalists took to mathematics, followed by the modern economic theorists. They did so not only because of the elegance and rigorous logic of mathematical models, but also in order to compete with their colleagues in the natural sciences, for whom mathematics was an absolute requirement. "Reigned" is maybe too lenient a description of the situation; it was more that the profession was gripped in a chokehold—so much so, in fact, that the often highly stylized models—complete with unworkable axioms—had hardly any connection to real life.

A general rethink was brought about by Herbert Simon, who took the initial step in describing what was actually happening in the real world, rather than what theorists thought ought to happen in a perfect world. True models of the universe do not recognize the Earth as the center of the galaxy simply because religious precepts wanted it to be so; they portray the Sun as the actual center because observations made that evident. In much the same way, Simon put *homo sapiens* at the center and developed models around him instead of positing axioms that *homo economicus* was supposed to obey. Considering human beings' bounded skills, he developed satisficing as a descriptive theory of human decision-making, as covered in more detail in chapter 11.

The cudgel was taken up by Daniel Kahneman and Amos Tversky, who, after conducting innumerable experiments on biases and heuristics, proposed prospect theory to describe decision-making under risk, thus crowning psychology, if not as a new queen, then surely as the princess of economic theory. This does not mean that there was no mathematics in their models—only that the underlying models were not based solely on rigorous, inflexible axioms.

In the late twentieth and early twenty-first centuries, it was back to economics, but with a twist. Richard Thaler (figure 13.1), a financial economist deeply influenced by the works of Kahneman and Tversky, spawned what would become known as *behavioral economics*. This relatively new field of economics does not orient itself by mathematical models based on stringent axioms; rather, it takes into account human frailties like biases and lack of willpower while determining human beings' decision-making processes. Thaler was the bridge spanning the gulf between psychology and economics.

Not everybody would have predicted that Thaler would one day become a professor, and a celebrated one at that. In fact, in high school, he was a straight-B student. Born in northern New Jersey in 1945, he grew up as the oldest of three boys. His father was an actuary; his mother a former schoolteacher turned stay-at-home mom. Bored and seemingly incompetent in school, he nevertheless applied and was accepted to Case Western Reserve University. He hesitated between psychology and economics, eventually opting for the latter because it offered better job prospects. When the time came to find a job, however, he decided to pursue graduate studies to avoid a career in business. His first academic position was an assistant professorship at the University of Rochester.

For one year, Thaler taught cost-benefit analysis to undergrads before advancing to a full-time job, without tenure, at the Rochester Graduate School of Management. To while away his time, he started a collection of "stupid things people do"—that is, decisions and patterns of behavior that his economist colleagues would call anomalies or irrational conduct. For example, when asking subjects what they would pay for a vaccine that would cure them of a disease that poses a 0.1 percent chance of dying, while asking others what they would demand in order to participate in a research trial with a 0.1 percent chance of dying, he found that the first group was prepared to pay about $10,000 for the cure, whereas the second group would demand $1 million to take part in the experiment. Why would people demand, in order to assume a risk, a payment that is 100 times higher than what they would pay to avoid the very same risk?

Or take the following scenario: Ms. S and her spouse have received free tickets to a concert. Unfortunately, on that evening, it's pouring rain outside, and they decide to forgo the concert. After all, the tickets did not cost anything. Would their decision have been different if the tickets had not been free? For many people, the answer is yes; if they had paid for the tickets themselves, they would have braved the weather, even though the payment for the tickets was by then a "sunk cost." (See chapter 12: past expenditures should not be relevant for current decisions.)

Or this scenario: Several years previously, Mr. R had bought a case of good wine for about $5 a bottle. Now his wine merchant offers to buy the wine back for $100 a bottle. Mr. R. refuses, although he would never pay more than $35 for a bottle of wine. According to the *endowment effect* identified by Thaler, people value an item higher if it already belongs to them than if they have to purchase it now.

Or this: Mr. H mows his own lawn. His neighbor's son would mow it for $8. But Mr. H wouldn't mow his neighbor's same-sized lawn, even for $20. By rejecting the boy's offer, he apparently ignored the "opportunity cost" of his labor (i.e., the more profitable things he could have done with his time).

FIGURE 13.1: Richard Thaler.

© Nobel Media AB Alexander Mahmoud

Or this: Ms. T enters a casino, gambles and wins $200. She continues gambling and loses all her winnings. Even though she just lost $200, she may not feel much pain because, in what Thaler termed *mental accounting*, she considers the $200 that she lost as belonging to the casino anyway.

To classical economists, all of these scenarios—and many more—seemed erroneous, irrational, or just plain dumb. According to their elegant mathematical models, such behavior must not occur, and if it does, it is only because some ignoramus does not know what he or she should be doing. The absurd behavior would represent no more than random noise, the economists maintained, and should wash out on average. But this view is wrong. With evidence mounting that most people act in such a fashion, the phenomenon could no longer be ignored. So, what was going on?

Apparently, people value an item that belongs to them more highly than the same item that they do not possess. And sunk costs are given consideration instead of being ignored, while opportunity costs are ignored instead of being treated like out-of-pocket expenses. The root cause of the anomalies that Thaler identified seemed to lie in the sharp differences between the buying and selling prices for the very same goods and services.

While pondering such anomalies—the sunk costs fallacy, opportunity costs, the endowment effect, mental accounting, and so on—without any real idea of where he was going with them, Thaler came across the works of Kahneman and Tversky. They were true eye-openers. These psychologists seemed to have the answers to all his inquiries.

As it happened, in 1977, the two Israelis were fellows at the Center for Advanced Study in the Behavioral Sciences at Stanford University. Thaler managed to secure a position at the Stanford branch of the National Bureau of Economic Research (NBER) for that year, and he soon became friends with Kahneman. What Thaler learned from him and discussed with him during that year would result in the paper "Towards a Positive Theory of Consumer Choice," published in the *Journal of Economic Behavior and Organization* in 1980. It would be considered the founding text of the emerging field of behavioral economics.

Unfortunately, his colleagues were unimpressed. In fact, nobody seemed in the least bit interested, and well-meaning mentors suggested that he go back to doing "real economics." Naturally, Thaler did not listen to their advice, and, as was also to be expected, Rochester did not grant him tenure. Fortunately, he obtained a position at Cornell University, where, for the next seventeen years, he developed his insights into a full-fledged theory.

In 1995, in spite of strenuous objection by some of the faculty there, he moved to the University of Chicago, the stronghold of rational economics. Recall that Chicago was home to the ultrarationalist Milton Friedman. However, by then, Thaler no longer needed to worry about his colleagues' derision: Behavioral economics had come into its own.

In 2015, this iconoclast, who had provided the passage between the economic and psychological analyses of individual decision-making, was elected president of the American Economic Association. Two years later, Thaler was awarded the Nobel Prize in Economics "for his contributions to behavioural economics." The curt announcement was expanded in the press release that followed. The laureate "incorporated psychologically realistic assumptions into analyses of economic decision-making," the full announcement read, and "has shown how these human traits systematically affect individual decisions as well as market outcomes."[2] By demonstrating that people are predictably irrational, Thaler was able to move economics toward a more realistic understanding of human behavior. The resulting insights have implications for public policy.

Thaler began his famous paper, "Towards a Positive Theory of Consumer Choice," with a nod to traditional models. "The economic theory of the consumer is a combination of positive and normative theories. Since it is based on a rational maximizing model, it describes how consumers should choose, but it is alleged to also describe how they do choose." And that was a problem, he argued, because "in certain well-defined situations, many consumers act in a manner that is inconsistent with economic theory. In these situations, economic theory will make systematic errors in predicting behavior."[3] It is Kahneman and Tversky's prospect theory that should be used as the descriptive theory.

Thaler then discusses several modes of behavior that would be classed as anomalies by traditional economists, but that he calls "economic mental illusions." In his terminology, these are classes of problems where consumers are particularly likely to deviate from the predictions of the normative model: underweighting of opportunity costs, failure to ignore sunk costs, suboptimal search behavior, choosing not to choose, and lack of self-control.

It was a total repudiation of received wisdom. "I agree with Friedman and Savage that positive theories should be evaluated on the basis of their ability to predict behavior. In my judgment, for the classes of problems discussed in this paper, economic theory fails this test." As a case in point, Thaler takes issue with Friedman and Savage's example of billiards players. Recall their argument that even though billiards players do not process mathematical formulas about the physics of incompressible spheres in their heads, they behave *as if* they did. Granted, says Thaler, that may be so for the very best: Friedman and Savage's mathematical model would be a good predictor of the behavior of an expert billiards player. Because the vast majority of billiards players are not experts, however, "it is instructive to consider how one might build models of two non-experts." Novice and intermediate billiards players would plan and execute their shots quite differently than the experts. They would play billiards using... yes, rules of thumb and heuristics. And, yes again, given their limited abilities, they would be totally rational in doing so.[4]

Similarly, nonexperts in economic theory—and the vast majority of consumers are nonexperts in data collection, probability theory, and computation—also do *not* behave as experts would. Still, within the bounds of their bounded rationality, they are quite rational.

"What I have argued," Thaler concludes in his paper, "is that the orthodox economic model of consumer behavior is, in essence, a model of robot-like experts. As such, it does a poor job of predicting the behavior of the average consumer. This is not because the average consumer is dumb, but rather that he does not spend all of his time thinking about how to make decisions."[5]

It is vital to note that the predilections of consumers to indulge in mental shortcuts, biases, and heuristics may be exploited for nefarious purposes. Casinos are notorious for making people believe that the odds of winning at their slot machines are better than they are; con artists abuse their victims' credulity; and climate change deniers confuse one-shot events with statistical evidence.

On the other hand, the very same predilections may be used to nudge people toward actions for their own good. Together with the Harvard law professor Cass Sunstein, Thaler wrote a book appropriately entitled *Nudge*. As the subtitle points out, the book is about *Improving Decisions About Health, Wealth, and Happiness*. By phrasing choice alternatives differently, or by switching the default option, people can be gently nudged toward improved decisions. A school cafeteria, for example, might try to push kids toward eating a good diet by putting the healthiest foods at the front, where they have a better chance of being chosen. An employer may increase participation in the corporation's sports program by forcing employees to opt out instead of inviting them to opt in.

Extremely confident in their outlook, Thaler and Sunstein did not hold back. Institutions, including governments, should enlist the science of choosing, they declared, "to steer people's choices in directions that will improve their lives." The result would be more savings for retirement, better investments, less obesity, more charitable giving, a cleaner planet, an improved educational system, and on and on. "The crazy thing is thinking humans act logically all the time," Thaler asserts in the 2015 comedy-drama film *The Big Short*, in which he made a cameo appearance alongside pop star Selena Gomez. The two appeared in a casino scene, in which they explained the hot-hand fallacy.[6]

The qualifier *gently* notwithstanding, not everybody was taken by their argument. Clearly, their stance was *Father Knows Best*, and critics of big government resented that the authors advocated that governments meddle with the free will of its citizens. Thaler and Sunstein countered that they simply recommended a sort of "libertarian paternalism" and the British daily *The Guardian* described nudging as "beyond left and right—it uses rightwing means to achieve progressive ends."[7]

There was a reason that *The Guardian* weighed in. Prime Minister David Cameron's administration in the United Kingdom had set up a Behavioural Insights Team in 2010, dubbed a "nudge unit," to develop policies, and Thaler was the team's formal advisor. And in the United States, Sunstein was entrusted by President Barack Obama with the task of adopting ideas from behavioral economics and incorporating them into policy.

As a financial economist, the idea of monetizing the results of his research was not very far from Thaler's mind. Together with Russell Fuller, an investment manager and former chairman of the finance department at Washington State University, he cofounded Fuller & Thaler, an asset management company, under the motto "Investors Make Mistakes. We Look For Them."[8] There are two kinds of mistakes, their corporate statement tells us: Investors may overreact to bad news and panic, or they may underreact to good news by not paying sufficient attention to it. Both mistakes produce buying opportunities for investors steeped in the intricacies of behavioral economics.

Using principles of behavioral finance, Fuller & Thaler has about $9 billion under management. As indicated previously, its strategy capitalizes on the market's overreaction to negative information and underreaction to favorable information. Accordingly, it invests in stocks that are subject to negative news regarding their future prospects, and in companies that are showing signs of recovery after a sustained period of disappointing performance.

As a concerned citizen, Thaler did his part to use behavioral economics for the community. Apart from his role on the Behavioural Insight Team in the United Kingdom, he served as an informal adviser to President Obama's administration and to his successful reelection campaign in 2012.

Herbert Simon had proposed that human beings are boundedly rational; he replaced utility maximization with satisficing. Daniel Kahneman and Amos Tversky had shown what the bounds are "by exploring the systematic biases that separate the beliefs that people have and the choices they make from the optimal beliefs and choices assumed in rational-agent models."[9] Richard Thaler took what is, for the present, the final step: By applying Kahneman and Tversky's experimental findings to the anomalies of rational choice theory, he incorporated the psychology of decision-making into economic models of behavior.

What's next? Artificial intelligence? Robotic rationality? Time will tell . . .

NOTES

1. IT ALL BEGAN WITH A PARADOX

1. For more on the Bernoulli family, see Szpiro (2010a), chapter 17. The quote, "Counselors at law are …," is from Bernoulli 1709, 56.
2. Rémond de Montmort 1713, 401 (my translation). In what follows, the excerpts of the correspondence between Bernoulli and Montmort, and between Bernoulli and Cramer, are from Pulskamp 2013. Subsequent quotes are from this correspondence unless otherwise indicated in the notes.
3. Note that Bernoulli here signed the letter as "Bernoully."
4. For more on this correspondence, see Devlin 2008.
5. The Dutch astronomer Christiaan Huygens (1657) claimed that the fair price is equal to the mathematical expectation or expected value.
6. Cramer used a formula that gives the sum of an infinite geometric progression.
7. For once, the master mathematician erred. Nikolaus wrote that the truncated series sums to 2 rather than 2.5.
8. Nikolaus ignores the probability of losing half one's money. As we will see later in this discussion, Daniel will do better than that.
9. Bernoulli 1738 ("Specimen Theoriae Novae De Mensura Sortis," *Econometrica* [1954]), 24.
10. Bernoulli 1738, 24.
11. Bernoulli 1738, 25.
12. To translate back from utility to monetary values, one must use the inverse of the logarithmic function, which is the exponential function: $\exp(7.8) = 2{,}440$.
13. Bernoulli 1738, 33.

2. MORE IS BETTER...

1. The courtesan's alleged beauty never ceased to inspire. When Karl Marx, in his doctoral thesis of 1841, searched for a metaphor for the absurdity of an attempt to bring Epicurus's philosophy into accordance with the tenets of the Catholic Church, he expressed it thus: "It is as if one tried to put a nun's cloak around Lais's blossoming body."
2. Diogenes Laërtius, *Book II*, Chapter 8.
3. Diogenes Laërtius, X:128.
4. Watson 1895, 58.
5. Diogenes Laërtius, X:132.
6. Diogenes Laërtius, X:11.
7. Diogenes Laërtius, X:121.
8. By law, a conviction required two witnesses. Because there was only one person who pointed his finger, a certain Lord Howard of Escrick, George Jeffreys (the notorious "hanging judge") decided that Sidney's manuscript—it was no more than that, and would only be published posthumously—could serve in lieu of a second witness. *Scribere est agere* (writing is the same as acting), he ruled.
9. Locke 1690, chapter 7, section 78.
10. Locke, chap. 5, section 46.
11. Locke, chap. 5, section 47.
12. Locke, chap. 5, sections 49, 46, 48, and 50.
13. Locke, chap. 2, section 4.
14. Locke, chap. 9, sections 123 and 124.
15. Bentham 1800, 50.
16. Bentham 1776, preface.
17. "Utilitarianism had been so distinctly in the air for more than a generation before [Bentham] published his *Principles of Morals and Legislation*, that he could not possibly have failed very substantially to profit by the fact" (Albee 1902, 167).
18. Bentham 1816.
19. According to the Stanford Encyclopedia of Philosophy, Bentham formulated this epithet while listening to William Blackstone's lectures in 1763–64. See "Jeremy Bentham," https://plato.stanford.edu/entries/bentham/. Last accessed July 19, 2019.
20. This is most likely one of Bentham's contributions to John Lind's *Answer to the Declaration of the American Congress* (1776).
21. A problem may arise concerning compulsory military service. How does a citizen's absolute right to life accord with the government's authority to send him to war,

where he risks his life? Arguments can be made in terms of the greater good of the nation, or for mercenary or volunteer armies.
22. After all, this is why they are called *individuals* and not, say, *homogenials*.
23. Marx 1867, 432.
24. Jefferson 1816. And a good thing, too—du Pont's son, Éleuthère Irénée du Pont de Nemours, founded the industrial giant E. I. Du Pont.
25. Universal Declaration of Human Rights (1948), Article 25.
26. International Covenant on Economic, Social and Cultural Rights (1966), Article 7.

3. ... AT A DECREASING RATE

1. After all, that is why goods are called *goods* and not *bads*.
2. Aristotle, *Nicomachean Ethics*, Book IV, chapter 1.
3. This is often translated as "classes of goods," but I believe "sources of contentment" conveys in a better way what Aristotle meant.
4. Aristotle is actually silent on physical well-being, but it may be assumed that he would put no bounds on how healthy one would like to be.
5. Aristotle, *Politika*, Book VII, chapter I.
6. Bentham 1834, vol. 3:229.
7. Bentham 1834 3:230.
8. Bentham 1834 3:230.
9. Bentham 1834 1:305; 3:230.
10. Marx 1867, 432.
11. And, as we saw in chapter 1, insurance is one other consequence of Axioms 3 and 4.
12. Laplace's work did not suffice, however, to satisfy the theoreticians of the following centuries. It was only after the Russian mathematician Andrei Kolmogorov put probability theory on a sound axiomatic foundation that it was considered sufficiently rigorous. For more on this point, see chapter 10 in Szpiro (2011); for more on Laplace and his work in general, see chapter 7 in Szpiro (2010).
13. Laplace 1814, 27.
14. It turned out that the exact interest rate was irrelevant, so long as it was positive, and so were the survival probabilities, so long as they diminished with advancing age. Interestingly, Laplace used a logarithmic utility but then approximated $log(1 + c)$, where 1 designated total wealth and c a small fraction of total wealth, by c. This actually implies a linear utility for wealth. Thus, he effectively denied Bernoulli's main principle and the main tenet of his own chapter (at least locally) that the utility increases at a decreasing rate.

218   3. ... AT A DECREASING RATE

15. Laplace 1812, vol. 2, chap. 10: 488.
16. Rabl 1909, 86.
17. In Weber's time, an ounce (*Unze*) was 31 grams.
18. Weber 1846, 115.
19. This recalls a legend about a Japanese master swordmaker. The secret to the strength of his swords lay in the precise temperature of the water in which the evolving sword was hardened. Once, when he showed his workshop to a visiting competitor, the guest pretended to slip on the floor and, in doing so, dipped his hand into the water to test its temperature. With a lightning-strike of his sword, so the story goes, the master sliced off the visitor's hand before the sensation of temperature could reach the cheater's brain.
20. Actually, Weber expressed temperatures in degrees Réaumur, with zero degrees at freezing, and 80 degrees at the boiling point.
21. A lot is a unit of weight, used in Europe since the Middle-Ages, corresponding to 1/30th of a pound.
22. Kuntze 1892, 39.
23. Breitkopf und Härtel, founded in 1719 in Leipzig, is still in business; today it is considered the world's oldest music publishing house. Among its clients were Ludwig van Beethoven, Franz Liszt, Richard Wagner, Frédéric Chopin, and Johannes Brahms.
24. Fechner 1860, v, ix.
25. Research performed by the neuroscientist Stanislas Dehaene and his group with the indigenous people of Brazil, who had little exposure to modern mathematics, suggests that humans' innate concept of numbers is not linear, but logarithmic.

4. THE MARGINALIST TRIUMVIRATE

1. Roscoe was also an author of some note, who wrote not only learned historical treatises but also a children's classic, *The Butterfly's Ball and the Grasshopper's Feast* (London: J. Harris, 1808), which he wrote to amuse his children. During its first year of publication, it sold no fewer than 40,000 copies, and the book remained popular for decades after that.
2. Collison Black 1977a, vol. 3, 33.
3. Collison Black and Konekamp 1972, vol. 1, 191.
4. Collison Black 1973, vol. 2, 410.
5. Jevons 1866, 283.
6. Except where otherwise noted, the following quotes are from Jevons (1871), 1.

7. Jaffé 1935, 190.
8. Walras 1860.
9. Dockès 1996, xxvii.
10. Jaffé 1935, 195.
11. Cournot 1838, viii.
12. Walras 1874b, 80.
13. Walras 1874a, 22.
14. Walras 1874b, 81.
15. Walras 1889, as translated by Jaffé 1954, 117.
16. Jevons 1977b, 40
17. The quotes here are from Jevons 1977b, 40, 41.
18. Collison Black 1977b, 46.
19. Jaffé 1983, 26.
20. Jevons 1879, preface.
21. See Szpiro 2011.
22. Carl's son Karl (*sic*) made up for this by becoming a noted mathematician. He even published an important paper on the so-called Super-Petersburg Paradox, in which the gains at the ith toss grow sufficiently faster than 2^i, to make expected utility infinite. Karl, born in 1902, when his father was already sixty-two years old, was the son of Menger and his longtime companion, Hermione Andermann. The couple never got married, possibly because Andermann was either divorced or Jewish, and Catholics like Menger (and, of course, Prince Rudolf) would have had to marry in religious ceremonies. Menger sought and eventually got an act of legitimacy for his son Karl from Emperor Franz-Josef.
23. Fisher 1892, 120.
24. To infer laws about economic behavior from statistics is the realm of a subdiscipline called *econometrics*.
25. Both quotes are from Schmoller 1883, 251.
26. Schmoller 1884.

5. FORGOTTEN PRECURSORS

1. I will not speak here about Augustin Cournot, who had already inspired Walras, father and son, as discussed in chapter 4.
2. Dupuis 1844, 87.
3. Jevons 1879, xxviii.
4. Gossen 1854, v.

220   5. FORGOTTEN PRECURSORS

5. Consumption takes time, which limits the quantitative intake of any good per day. Thus, according to Karl Heinrich Rau (1792–1870), another little-known precursor, satiation (and diminishing marginal utility) may simply be the result of time constraints.
6. Jevons 1879, xxxv.
7. Jevons (1879) wrote that Gossen's "curves of utility are generally taken as straight lines" (xxxvi). In today's manner of speaking, this would cause a misunderstanding. What Jevons means is that marginal utility is a straight line (i.e., it decreases linearly). Total utility, then, is a concave curve.
8. Jevons 1879, xxxviii.
9. Euler was Swiss, not German, but the sentiment still holds. The quote is from Walras 1830, 87.
10. Other forgotten precursors also include Nicolas-François Canard (1750–1833), Claus Kröncke (1771–1843), Francesco Fuoco (1774–1841), Graf Georg von Buquoy (1781–1851), Johann Heinrich von Thünen (1783–1850), and Hans von Mangoldt (1824–1868).

6. BETTING ON ONE'S BELIEF

1. Forrester 2004, 12.
2. Indeed, in 1929, a research Fellow at Magdalen, where Ramsey's father was president, was expelled because "French Letters" (a euphemism for condoms) had been discovered beside his bed.
3. All quotes in this paragraph are from Galavotti (2006), 15.
4. Ramsey 1926, 157.
5. Keynes 1930, 153.
6. Ramsey 1926, 12.
7. Ramsey 1926, 12.
8. Ramsey 1926, 14.
9. This quote is from Ramsey 1926, 17. But how much simpler is Yogi Berra's advice: "When you come to a fork in the road, take it."
10. See Szpiro 2010b for problems that can arise if an assessment system is not consistent.
11. I use the word *occurrence* a bit loosely here. More precisely, Ramsey considers the truth of propositions, such as the belief that God exists or that the Riemann Conjecture is true.

12. Expressed more formally (where ~ denotes "not"), the laws are as follows:

   (1) Degree of belief in p, plus degree of belief in $\sim p = 1$.
   (2) Degree of belief in p given q, plus degree of belief in $\sim p$ given $q = 1$.
   (3) Degree of belief in (p and q) = degree of belief in p × degree of belief in q given that p occurred.
   (4) Degree of belief in (p and q), plus degree of belief in (p and $\sim q$) = degree of belief in p.

13. Note that the odds of a win by the Grasshoppers or the Butterflies play no role here—the gambles need not be "fair." The point is that the fourth law is violated, which produces a loss for the gambler, no matter what the odds of the wagers are.
14. Ramsey 1926, 22.
15. In formal terms, the following are Kolmogorov's definitions and axioms of probability theory:

    Let Ω be a nonempty set (which we will call "the universal set"). A field (or algebra) on Ω is a set F of subsets of Ω that has Ω as a member, and which is closed under complementation (with respect to Ω) and union. Let P be a function from F to the real numbers, obeying the following:

    Nonnegativity: $P(A) \geq 0$, for all $A \in F$.
    Normalization: $P(\Omega) = 1$.
    Finite additivity: $P(A \cup B) = P(A) + P(B)$ for all $A, B \in F$, such that $A \cap B = \emptyset$.

    P is called a *probability function*, and (Ω, F, P) a *probability space*.

7. GAMES ECONOMISTS PLAY

1. For more on this fascinating person's life, see Szpiro 2010b.
2. The quote is from von Neumann 1928, 295. It is regrettable that this hodgepodge of a phrase, *homo economicus*, a combination of Latin and Greek terms, has found its way into academic discourse. More correctly, it should be the pure Latin *homo parcus* (to distinguish from simple-minded *homo sapiens*), or the pure Greek, *anthropos oikonomikos*.
3. Morgenstern 1976, 805.
4. The Rockefeller Foundation had declared business cycles one of its main research areas in the 1920s and 1930s.

5. Morgenstern 1976, 808. Subsequent quotes, unless otherwise indicated in the notes, are to this edition.
6. Morgenstern, *Diaries*, March 4, 1923 (Duke University). Subsequent citations from the diary entries are to this source.
7. von Neumann and Morgenstern 1944, 16.
8. That statement seems intuitively incorrect, but it becomes obvious when translating degrees Fahrenheit into degrees Celsius. In this example, 20°F corresponds to about –7°C, but 40°F corresponds to about 4°C. Of course, the latter is not twice the former.
9. In what follows, I give a somewhat simplified version of the axioms than von Neumann and Morgenstern presented, which I hope will be easier to understand.
10. See Szpiro 2010b.
11. Recall the game rock, paper, scissors was mentioned earlier in this chapter. The game's endless boredom is based on the fact that rock, paper, and scissors also form a cycle.
12. This is an anecdote that is ascribed to the philosopher Sidney Morgenbesser from Columbia University (Szpiro 2010b).
13. It can be argued that in the presidential election in the United States in 2000, George W. Bush beat Al Gore, even though a majority may have preferred Gore, because Ralph Nader, who had no chance of winning at all, presented himself as a third-party candidate.
14. For an exploration of that topic, see Szpiro 2010b.
15. Morgenstern 1976, 809.
16. These quotes are from von Neumann and Morgenstern 1944, 617, 618, 627.
17. In terms of von Neumann and Morgenstern's theory, this means that if $f(x)$ and $g(x)$ are two utility functions for the same person, then $g(x) = a + bf(x)$ for any a and any positive b. Also, if $f(x)$ is a person's utility function, then any *other* function $h(x)$ that can be expressed as $h(x) = c + df(x)$ may also serve as his or her utility function.
18. Actually, everybody can pick one option from infinitely many utility functions because the factor and the constant can be chosen freely.
19. Planck 1948, 22.
20. Lissner 1946, 1.
21. Hurwicz 1945, 924; Marschak 1946, 98; Simon 1945, 559; Stone 1948, 200; Copeland 1945, 498.

8. WOBBLY CURVES

1. Beregszasz is today's Berehovo in Ukraine. This author's mother also hails from Beregszasz, where some of the Friedman family were neighbors.

2. In his PhD dissertation, he showed that the medical profession's monopoly powers had raised substantially the incomes of physicians relative to that of dentists.
3. Actually, the official name of this prize is "The Sveriges Riksbank Prize in Economic Sciences in Memory of Alfred Nobel."
4. Sampson and Spencer 1999, 128.
5. After all, in chapter 2, we showed that more is better than less.
6. Both quotes are from Friedman and Savage 1948, 286. Subsequent citations are from this edition unless otherwise noted.
7. Markowitz 1952a, 153. Subsequent citations are from this edition unless otherwise noted.
8. This predates by several decades the theories of behavioral economics, which I describe in chapters 12 and 13.
9. Markowitz 1952b.
10. Nowadays, hardly any scientific journal would accept a paper whose data analysis was based on a quaint "the typical answers (of my middle-income acquaintances) to these questions...." In order to publish academic papers, large sample sizes, T-statistics, and control groups, among other elements, are required.
11. Markowitz 1990.

9. COMPARING THE INCOMPARABLE

1. In 1960, the standardization of the meter was changed to the wavelength of the radiation of the krypton atom. The specification of the kilogram was changed as of May 20, 2019, and is now defined by the Planck constant, a concept in quantum mechanics.
2. To obtain degrees Celsius, one simply deducts 32 from the degrees Fahrenheit, and then divides the remainder by 1.8.
3. Parts of this chapter are adapted from Szpiro 2003 and Szpiro 2010b.
4. Starr (2008), 233.
5. This quote is from Arrow 1951, 59; see also Szpiro 2010b.
6. The majority rule can result in so-called Condorcet cycles. See Szpiro 2010b.
7. Pratt 1964, 122.
8. In mathematics, different definitions of curvature exist. To simplify this, I here take the second derivative of the utility function, $U''(wealth)$, as a proxy for the graph's curvature. Note that while the fact that $U'(wealth)$ is positive indicates that people prefer more to less, the fact that $U''(wealth)$ is negative indicates a dislike for variance (i.e., risk).

9. The mathematically inclined reader may note that even though utility functions are defined only up to linear transformations, risk aversions are invariant under linear transformations.
10. For example, Szpiro 1986.
11. As a postscript to this chapter, let me mention a mathematical expression called the *Schwarzian derivative of a function*: $S(W) = U'''(W)/U'(W) - 3/2[U''(W)/U'(W)]2$. It includes not only Pratt and Arrow's degree of risk aversion but also the term $U'''(W)$, which can be interpreted as an inclination or a dislike for the skewness of the wealth distribution. Many years ago, at a symposium in Jerusalem, I had the occasion to corner Kenneth Arrow and ask him about that. He knew about the Schwarzian derivative but told me that he had never been able to figure out its significance in economics. (See also Szpiro 1988.)

10. MORE PARADOXES

1. Martin 2010.
2. Allais 1988, 3.
3. Samuelson 1986, 84.
4. Because Poincaré was French, Allais may have let himself be ensnared in this controversy by patriotic, if not to say nationalistic, sentiments. In any case, the charge of plagiarism has been disproven (see Szpiro 2007). Also, there is a bit of a problem here: On the one hand, Allais claimed that Einstein plagiarized the theory of relativity, while on the other hand, his experiment would disprove this very theory.
5. I met Allais in June 1988, at the Fourth International Conference on Foundations and Applications of Utility, Risk, and Decision Theory in Budapest. I was quite flattered that he came to listen to my talk at the conference, and that later, he talked to me, one on one, at length about his life and achievements. A few days after the conference, a package arrived at my home with a host of reprints of his scientific papers. I dutifully wrote a report about the conference for my newspaper, the Swiss daily *Neue Zürcher Zeitung*, of course mentioning Allais's importance to economic theory. Only later did it dawn on me that the reason he sought me out was neither the quality of my own (very modest) scientific work, nor my charming personality, but rather the fact that I was a journalist. It was just four months before the Nobel Prize would be bestowed on him.
6. Allais used French francs.
7. Allais 1953, 527.
8. Wells 2001, 49.

9. Ellsberg 2006.
10. United Press International 1971.
11. Ellsberg phrased the questions a little differently. The way in which I present them here brings out the "independence of irrelevant alternatives" axiom in a manner that is a bit simpler to understand. The paradox can also be set up for those who answered "blue."
12. Ellsberg 1962, 651.
13. The American economist Frank Knight differentiated between risk, where the odds are known (e.g., sixty balls out of ninety), and uncertainty, where they are not (between thirty and sixty balls out of ninety).

11. GOOD ENOUGH

1. Simon 1955, 99.
2. Simon 1978.
3. It moved to Yale after a dispute with the economics department at Chicago and was renamed the Cowles Foundation for Research in Economics.
4. Both quotes are from Simon 1955, 99.
5. In fact, this author, as a student of mathematics, first encountered economics in Gérard Debreu's *Theory of Value*, a treatise that presents economics purely as an example of mathematical modeling.
6. Conlisk 1996, 670.
7. Selten 2001, 14.
8. Merton and Scholes received the 1997 Nobel Prize in Economics; for more information, see Szpiro 2011.
9. Simon 1955, 101.
10. Note, however, that by positing the diminishing marginal utility of wealth, Bernoulli had actually already incorporated the human mind (i.e., psychology) into the study of economic decision-making.
11. Simon 1955, 99.

12. SUNK COSTS, GAMBLER'S FALLACY, AND OTHER ERRORS

1. Both quotes are from Tversky and Kahneman 1974, 1124.
2. Lewis 2016.

3. Kahneman shared the prize with Vernon Smith. (Tversky could not be honored since Nobel prizes are not awarded posthumously.) The quote is from NobelPrize.org 2017b.
4. Kahneman 2002.
5. Apparently, an average is simpler for the human mind to conceive than a sum.
6. I am simplifying the experimental setup to make it easier to understand.
7. We can see this violation all the time in advertising. A sexy supermodel on the hood of a sports car in an advertisement, for example, should be ignored, as she has nothing to do with the car's quality. But apparently—or at least according to the expectations of advertisers—the addition of the model makes men want to purchase that model (of the car, that is!).
8. These quotes are from Tversky and Kahneman 1974, 1128.
9. Tversky and Kahneman 1974, 1128.
10. One evening, on a dark street, Peter encounters Paul, who is looking for something next to the streetlamp. "I lost my car keys in the parking lot," says Paul. "So why are you looking here, not there?" asks Peter. "This is where the light is," comes the answer.
11. Tversky and Kahneman 1974, 1128.
12. Do not—repeat: *do not, ever,* apply regression to the mean to coin tosses or spins of a roulette wheel. If you do this, you will have fallen for the gambler's fallacy, described later in this chapter.
13. Harry Markowitz had already made a similar point. (See chapter 8.)
14. The record for the longest string of one color at roulette was thirty-two consecutive reds at an American casino in 1943. The probability of such a sequence is 0.000000023 percent. In the St. Petersburg Paradox, this would have resulted in a payout of nearly $4.3 billion. However, note that on the thirty-first spin, the gambler would have had to wager more than $2.1 billion.
15. The word could also refer to other kinds of decision problems, like the choice of a spouse, the purchase of a home, educational possibilities or, more broadly, general welfare. The quotes are from Kahneman and Tversky (1979), 263.
13. Note that von Neumann and Morgenstern, as well as other game theorists, would have no problem with operations 2, 3, 4, and 6 of the editing process. They are simplification procedures that conform with all axioms of rational decision-making.
16. Friedman and Savage had regarded utility as a function of monetary income, with zero income implied as zero utility. But Markowitz actually had already considered current wealth as the reference point.
17. Kahneman and Tversky (1979), 279.
18. Kahneman and Tversky (1979), 280.

19. If we denote the subjective value of the payout as $v(\text{amount in dollars})$, then the two decision problems that make up the Allais Paradox can be written as follows (you may want to refer back to chapter 9 before continuing):

   (A) $v(1{,}000) > \pi(0.10) \cdot v(5 \text{ million}) + \pi(0.89) \cdot v(1 \text{ million}) + \pi(0.01) \cdot v(0)$
   and
   (B) $\pi(0.10) \cdot v(5 \text{ million}) + \pi(0.90) \cdot v(0) > \pi(0.11) \cdot v(1 \text{ million}) + \pi(0.89) \cdot v(0)$.

   Because $v(0) = 0$, (A) and (B) can be rewritten as

   (A') $[1 - \pi(0.89)] \cdot v(1 \text{ million}) > \pi(0.10) \cdot v(5 \text{ million})$
   (B') $\pi(0.10) \cdot v(5 \text{ million}) > \pi(0.11) \cdot v(1 \text{ million})$.

   Hence, $[1-\pi(0.89)] \cdot v(1 \text{ million}) > \pi(0.11) \cdot v(1 \text{ million})$ or, written differently, $\pi(0.89) + \pi(0.11) < 1$.

20. The experiment's result says that $\pi(0.001) \cdot v(6{,}000) > \pi(0.002) \cdot v(3{,}000)$. Hence, $\pi(0.002)/\pi(0.001) < v(6{,}000)/v(3{,}000)$. Because the marginal value of money—like marginal utility—decreases, $v(6{,}000)/v(3{,}000) < 2$. Hence, $\pi(0.002)/\pi(0.001) < 2$ or $\pi(0.002) < 2 \cdot \pi(0.001)$. This is the definition of subadditivity.

21. Consider these two decision problems:

   (A) Receive $4,000 with probability of 80 percent, or $3,000 for sure.
   (B) Receive $4,000 with probability of 20 percent, or $3,000 with probability of 25 percent.

   In gamble (A), most people opt for the sure thing, and in gamble (B), most people opt for the first of the two alternatives. Note, however, that problem (B) is the same as problem (A)—it is simply scaled down by three-quarters. This means that

   (A') $\pi(1) \cdot v(3{,}000) > \pi(0.80) \cdot v(4{,}000)$, hence $\pi(0.80)/\pi(1) < v(3000)/v(4000)$
   and
   (B') $\pi(0.20) \cdot v(4{,}000) > \pi(.25) \cdot v(3{,}000)$, hence $v(3000)/v(4000) < \pi(0.20)/\pi(0.25)$.

   Therefore, $\pi(0.80)/\pi(1.00) < \pi(0.20)/\pi(0.25)$, or in general terms, $\pi(p)/\pi(q) < \pi(\alpha p)/\pi(\alpha q)$, where α is a constant.
   This is satisfied if $Ln(\pi(p))$ is a convex function of $Ln(p)$.

22. One must distinguish *overweighting*, which refers to a property of decision weights, from the *overestimation* that is commonly found in the assessment of the probability of rare events. Overestimation is the consequence of some heuristics.

23. $\pi(0.001) \cdot v(5,000) > v(5)$. Hence, $\pi(0.001) > v(5)/v(5,000)$ and because marginal value decreases, $v(5,000) < 100 \cdot v(5)$. Thus, $\pi(0.001) > 0.001$.
24. But there is still another model of decision-making, called *ecological rationality* (from *ecology*, the science of how organisms interact with each other and with the environment). Pioneered by Vernon Smith (who shared the Nobel Prize in Economics in 2002 with Daniel Kahneman) and Gerd Gigerenzer of Germany, it says that heuristics are ecologically rational if they exploit structures of information in the environment.

    For example, when asking Europeans what the most populous metropolitan area is in North America, they may say New York City, simply because they often see or hear its name in the media. They are wrong—Mexico City has a larger metropolitan population—but because the number of media mentions of that city does not correlate with its population size, this heuristic is not at all unreasonable. After all, New York City comes second.

13. ERRONEOUS, IRRATIONAL, OR PLAIN DUMB?

1. Smith 1776, *Book IV*, Chapter V, 534.
2. These quotes are from NobelPrize.org 2017a, 2017b. Amazingly, Michael Lewis had foreseen, or rather surmised, in his bestseller *The Undoing Project*, that Thaler would be a Nobel recipient. On page 278, he wrote that Thaler "wasn't... anyone's idea of a future Nobel Prize winner." The book was published in the fall of 2017, several weeks *before* the Nobel prizes were announced!
3. Thaler 1980, 39.
4. These quotes are from Thaler 1980, 57, 58.
5. Thaler 1980, 58.
6. Thaler and Sunstein 2008, 5; McKay 2015.
7. Chakrabortty 2008. The article remains silent about the fact that the same means could be used to achieve disreputable ends too.
8. Fuller & Thaler. https://www.fullerthaler.com/about.
9. Kahneman 2003, 1449.

BIBLIOGRAPHY

Albee, Ernest. 1902. *A History of English Utilitarianism*, London: Swan Sonnenschein & Co.
Allais, Maurice. 1943. *À la Recherche d'une Discipline Économique* [In quest of an economic discipline]. Paris: Ateliers Industria.
———. 1947. *Économie et Interêt* [Economy and interest]. Paris: Imprimerie Nationale.
———. 1953. "Le Comportement de l'Homme Rationnel devant le Risque: Critique des Postulats et Axiomes de l'Ecole Americaine." (Behavior of rational man when faced with risk: critique of the postulates and axioms of the American School.) *Econometrica* 21 (4): 503–546.
———. 1988. "An Outline of My Main Contributions to Economic Science. Nobel Lecture, December 9, 1988." *American Economic Review* 87 (6): 3–12.
Aristotle. 1906. *Nicomachean Ethics* [c.340 BCE]. London: Kegan Paul. https://www.stmarys-ca.edu/sites/default/files/attachments/files/Nicomachean_Ethics_0.pdf.
———. *Politika* [c.350 BCE]. Available online at http://classics.mit.edu/Aristotle/politics.html.
Arrow, Kenneth. 1951. *Social Choice and Individual Values*. New York: Wiley.
———. 1965. *Aspects of the Theory of Risk Bearing*. Helsinki: Yrjö Jahnssonin.
Ashley-Cooper, Anthony, 3rd Earl of Shaftesbury. 1711. *Characteristicks of Men, Manners, Opinion, and Times*, 3 vols. (Reprinted by Kessinger Publishing, Whitefish NY, 2010.)
Bentham, Jeremy. 1843. *The Works of Jeremy Bentham, Published Under the Supervision of His Executor, John Bowring*, 11 vols. Edinburgh: Tait. https://oll.libertyfund.org/titles/bentham-works-of-jeremy-bentham-11-vols

———. "Anarchical Fallacies," in *Works*, 1843, Vol. II.
———. "Fragment on Government," in *Works*, 1843, Vol. I.
———. "A Manual of Political Economy," in *Works*, 1843, Vol. III.
———. *Pannomial Fragments*, in *Works*, 1843, Vol. III.
———. *Principles of the Civil Code*, in *Works*, 1843, Vol. I.
———. "Principles of Morals and Legislation," in *Works*, 1843, Vol. I.
———. "Short Review of the Declaration," in *An Answer to the Declaration of Independence*, ed. John Lind, 1776. https://archive.org/details/cihm_20519/page/n17, 119–132.
Bernoulli, Daniel. 1954. "Specimen Theoriae Novae De Mensura Sortis" [Exposition of a new theory for the measurement of chance], 1738; published in English in *Econometrica* 22: 23–36.
Bernoulli, Nikolaus. 1709. *De Usu Artis Conjectandi in Jure* [On the use of the technique of conjecturing on matters of law], https://books.google.co.il/books?id=4xd8nQAACAAJ&printsec=frontcover&source=gbs_ge_summary_r&cad=0#v=onepage&q&f=false English translation https://www.cs.xu.edu/math/Sources/NBernoulli/de_usu_artis.pdf.
Bordas, Louis. 1847. "De la mesure de l'utilité des travaux publics." *Annales des Ponts et Chaussées* 2nd ser. 249–284.
Chakrabortty, Aditya. 2008. "From Obama to Cameron, Why Do So Many Politicians Want a Piece of Richard Thaler?" *Guardian*, July 12, https://www.theguardian.com/politics/2008/jul/12/economy.conservatives.
Collison Black, R. D., ed. 1973. *Correspondence, 1850–1862*. Vol. 2 of *Papers and Correspondence of William Stanley Jevons*. London: Palgrave Macmillan.
———. 1977a. *Correspondence, 1863–1872*. Vol. 3 of *Papers and Correspondence of William Stanley Jevons*. London: Palgrave Macmillan.
———. 1977b. *Correspondence, 1873–1878*. Vol. 4 of *Papers and Correspondence of William Stanley Jevons*. London: Palgrave Macmillan.
Collison Black, R. D., and Rosamond Konekamp, eds. 1972. *Biography and Personal Journals*. Vol. 1 of *Papers and Correspondence of William Stanley Jevons*. London: Palgrave Macmillan.
Conlisk, John. 1996. "Why Bounded Rationality?" *Journal of Economic Literature* 34 (2): 669–700.
Copeland, Arthur. 1945. "Review: John von Neumann and Oskar Morgenstern, *Theory of Games and Economic Behavior*." *Bulletin of the American Mathematical Society* 51: 498–504.
Cournot, Antoine. 1838. *Recherches sur les principes mathématiques de la théorie des richesses*. Paris: Hachette.

Debreu, Gérard. 1972. *Theory of Value: An Axiomatic Analysis of Economic Equilibrium.* Cowles Foundation Monographs Series. New Haven, CT: Yale University Press.

Devlin, Keith. 2008. *The Unfinished Game: Pascal, Fermat, and the Seventeenth-Century Letter That Made the World Modern.* New York: Basic Books.

Diogenes Laërtius. n.d. *Lives of the Eminent Philosophers*, 180–270 CE. https://en.wikisource.org/wiki/Lives_of_the_Eminent_Philosophers

Dockès, Pierre. 1996. *La société n'est pas un pique-nique: Léon Walras et l'économie sociale.* [Society is no picknick: Léon Walras and social economics], Paris: Économica.

Dupuit. Jules. 1844. "De la mesure de l'utilité des travaux publics" [On measuring the usefulness of public works]. *Annales des Ponts et Chaussees.* Reprinted in *Revue Française d'Économie* 10, no. 2 (1995): 55–94, http://bibliotecadigital.econ.uba.ar/download/Pe/181746.pdf.

———. 1849. "De l'influence des péages sur l'utilité des voies de communication." [About the influence of tolls on the utility of transportation routes], *Annales des Ponts et Chaussées*, 2nd ser.: 170–248.

Ellsberg, Daniel. 1961. "Risk, Ambiguity, and the Savage Axioms." *Quarterly Journal of Economcs* 75: 643–669.

———. 2001. *Risk, Ambiguity, and Decision.* New York: Garland.

———. 2006. "Extended Biography." http://www.ellsberg.net/bio/extended-biography/.

Epicurus. n.d. *Letter to Menoeceus*, https://users.manchester.edu/Facstaff/SSNaragon/Online/texts/316/Epicurus,%20LetterMenoeceus.pdf.

Epicurus. n.d. *Vatican Sayings*, (14th century). http://epicurus.net/en/vatican.html.

Fechner, Gustav Theodor. 1860. *Elemente der Psychophysik* [Elements of psychophysics]. Leipzig, Germany: Breitkopf & Härtel.

Fisher, Irving. 1961. *Mathematical Investigations in the Theory of Value and Prices.* Originally published in 1892. New York: A. M. Kelley.

Forrester, John. 2004. "Freud in Cambridge." *Critical Quarterly* 46 (2): 12.

Friedman, Milton, and L. J. Savage. 1948. "The Utility Analysis of Choices Involving Risk." *Journal of Political Economy* 56 (4): 279–304.

———. 1952. "The Expected-Utility Hypothesis and the Measurability of Utility." *Journal of Political Economy* 60 (6): 463–474.

Fuller & Thaler Asset Management. n.d. About the Company, https://www.fullerthaler.com/about.

Galavotti Maria C., ed. 2006. *Cambridge and Vienna: Frank P. Ramsey and the Vienna Circle.* Heidelberg, Germany: Springer.

Gossen, Hermann Heinrich. 1854. *Entwickelung der Gesetze des menschlichen Verkehrs, und der daraus fliessenden Regeln für menschliches Handeln* [Development of the laws of human commerce, and of the resulting rules of human action]. Braunschweig, Germany: Friedrich Vieweg und Sohn.

Hurwicz, Leonard. 1945. "The Theory of Economic Behavior." *American Economic Review* 35 (5): 909–925.

Huygens, Christiaan. 1657. "De ratiociniis in aleae ludo" [On Reasoning in Games of Chance]. In Frans van Schooten, *Exercitationum Mathematicarum*. Leyden, Netherlands: Johannis Elsevirii, 521–534.

Jaffé, William. 1935. "Unpublished Papers and Letters of Léon Walras." *Journal of Political Economy* 43 (2): 187–207.

———. 1954. Translation of "Elements of Pure Economics", London: George Allen and Unwin.

———. 1983. *William Jaffé's Essays on Walras*, ed. Donald A. Walker. Cambridge: Cambridge University Press.

Jefferson, Thomas. 1816. *Letter to P. S. Dupont de Nemours*, April 24. http://www.let.rug.nl/usa/presidents/thomas-jefferson/letters-of-thomas-jefferson/jefl243.php.

Jevons, Stanley. 1863. *A Serious Fall in the Value of Gold Ascertained, and Its Social Effects Set Forth, with Two Diagrams*. London: Edward Stanford.

———. 1865. *The Coal Question: An Inquiry Concerning the Progress of the Nation and the Probable Exhaustion of Our Coal Mines*. London and Cambridge: Macmillan.

———. 1866. "A Brief Account of a General Mathematical Theory of Political Economy." *Journal of the Royal Statistical Society* 29: 282–287.

———. 1871. *The Theory of Political Economy*, 2nd ed. (5th ed., 1879; reprinted New York: August N. Kelley, 1957, 1965). https://archive.org/stream/JevonsW.S1871TheTheoryOfPoliticalEconomy/Jevons%2C%20W.%20S%20%281871%29%2C%20The%20Theory%20of%20Political%20Economy_djvu.txt

Kahneman, Daniel. 2002. *Maps of Bounded Rationality: A Perspective on Intuitive Judgment and Choice*, Nobel Prize Lecture, https://www.nobelprize.org/uploads/2018/06/kahnemann-lecture.pdf.

———. 2003. "Maps of Bounded Rationality: Psychology for Behavioral Economics." *American Economic Review* 93 (5): 1449–1475.

Kahneman, Daniel, and Amos Tversky. 1972. "Subjective Probability: A Judgment of Representativeness." *Cognitive Psychology* 3: 430–454.

———. 1979. "Prospect Theory: An Analysis of Decision Under Risk." *Econometrica* 47 (2): 263–292.

Keynes, John Maynard. 1921. *Treatise on Probability*. London: Macmillan.

———. 1930. "F. P. Ramsey." *Economic Journal* 40: 153–154.

Klausinger, Hansjörg. 2013. *Academic Anti-Semitism and the Austrian School: Vienna, 1918–1945*. Department of Economics, Wirtschaftsuniversität Wien, Working Paper 155.

Kolmogorov, Andrey Nicolayevich. 1933 [1950]. *Foundations of the Theory of Probability*, 2nd English ed., New York: 1950.

Kuntze, Johannes Emil. 1982. *Gustav Theodor Fechner (Dr. Mises) Ein deutsches Gelehrtenleben*. Leipzig, Germany: Breitkopf & Härtel.

Laplace, Pierre-Simon. 1814. *Théorie analytique des probabilités*, 2nd ed. Paris: Courcier.

———. 1840. *Essai philosophique sur les probabilités*, 6th ed. [Philosophical essay on probability]. Paris: Bachelier, 1840 (Originally published in 1814).

Leonard, Robert. 2010. *Von Neumann, Morgenstern, and the Creation of Game Theory*, Cambridge: Cambridge University Press.

Lewis, Michael. 2016. *The Undoing Project: A Friendship That Changed Our Minds*. New York: Norton.

Lissner, Will. 1946. "Mathematical Theory of Poker Is Applied to Business Problems." *New York Times*, March 10.

Locke, John. 1690. "An Essay Concerning the True Original Extent and End of Civil Government," in *Two Treatises on Government*. (Reprinted Boston: Edes and Gill, 1773.) http://www.bookwolf.com/newsite_0920_No_Use/Wolf/pdf/JohnLocke-essayConcerningTheTrueOriginalExtenet.pdf.

Markowitz, Harry. 1952a. "The Utility of Wealth." *Journal of Political Economy* 60 (2): 151–158.

———.1952b. "Portfolio Selection." *Journal of Finance* 7 (1): 77–91.

———. 1990. *Nobel Prize Lecture*, https://www.nobelprize.org/prizes/economic-sciences/1990/markowitz/biographical/.

Marschak, J. 1946. "Neumann's and Morgenstern's New Approach to Static Economics." *Journal of Political Economy* 54 (2): 97–115.

Martin, Douglas. 2010. "Maurice Allais, Nobel Winner, Dies at 99." *New York Times*, October 11.

Marx, Karl. 1867. *Das Kapital, Volume I*., https://www.marxists.org/archive/marx/works/download/pdf/Capital-Volume-I.pdf.

McKay, Adam. 2015. *The Big Short*. Regency Enterprises and Plan B Entertainment.

Menger, Carl. 1871. *Grundsätze der Volkswirthschaftslehre* [Principles of economics]. Vienna: Wilhelm Braumüller.

———. 1883. *Untersuchungen über die Methode der Socialwissenschaften* [Inquiries into the methods of the social sciences].Leipzig, Germany: Duncker und Humblot.

———. 1884. *Die Irrthümer des Historismus in der deutschen Nationalökonomie* [The errors of historicism in German political economy]. Vienna: Alfred Hölder.

Morgenstern, Oskar. 1976. "The Collaboration Between Oskar Morgenstern and John von Neumann on the Theory of Games." *Journal of Economic Literature* 14: 805–816.

———. n.d. *Oskar Morgenstern Papers*, Rubenstein Rare Book and Manuscript Library, Duke University.

NobelPrize.org. 2017a. "Press Release: The Prize in Economic Sciences 2017." https://www.nobelprize.org/prizes/economic-sciences/2017/press-release/.

———. 2017b. "The Prize in Economic Sciences 2017." https://www.nobelprize.org/prizes/economic-sciences/2017/summary/.

Planck, Max. 1948. *Wissenschaftliche Selbstbiographie*. Leipzig, Germany: Johann Ambrosius Barth Verlag.

Pratt, John. 1956. *Some Results in the Decision Theory of One-Parameter Multivariate Polya-Type Distributions*. PhD thesis, Stanford University, 1956.

———. 1964. "Risk Aversion in the Small and in the Large." *Econometrica* 32 (1/2): 122–136.

———. 1995. *Introduction to Statistical Decision Theory*. Cambridge, MA: MIT Press.

Programme on Women's Economic, Social, and Cultural Rights. 2015. *Human Rights for All: International Covenant on Economic, Social and Cultural Rights (1966)*. http://www.pwescr.org/PWESCR_Handbook_on_ESCR.pdf.

Pulskamp, Richard J., trans. 2013. "Correspondence of Nicolas Bernoulli Concerning the St. Petersburg Game," in *Die Werke von Jakob Bernoulli*, Vol. 3, http://cerebro.xu.edu/math/Sources/NBernoulli/correspondence_petersburg_game.pdf.

Rabl, Carl. 1909. *Geschichte der Anatomie an der Universität Leipzig*, https://archive.org/stream/geschichtederana00rabluoft/geschichtederana00rabluoft_djvu.txt.

Ramsey, Frank Plimpton. 1926. "Truth and Probability," in *The Foundations of Mathematics and Other Logical Essays*, ed. R. B. Braithwaite. New York: Harcourt, Brace, and Company, 1926, 156–198.

———. 1927. "A Contribution to the Theory of Taxation." *The Economic Journal* 37 (145): 47–61.

———. 1928. "A Mathematical Theory of Saving." *The Economic Journal* 38 (152): 543–559.

Rémond de Montmort, Pierre. 1713. *Essay d'analyse sur les jeux de hazard*, Jacques Quillau Imprimeur-juré libraire de l'Université, https://books.google.co.il/books?id=e6EXbosJmt8C&printsec=frontcover&source=gbs_ge_summary_r&cad=0#v=onepage&q&f=false.

Roscoe, William. 1802. *The Butterfly's Ball, and the Grasshopper's Feast*. London: J. Harris.

Royal Swedish Academy of Sciences. 2002. "Press Release." https://www.nobelprize.org/prizes/economic-sciences/2002/press-release/.

Sampson, A. R., and B. Spencer. 1999. "A Conversation with I Richard Savage." *Statistical Science* 14 (1): 126–148.

Samuelson, Paul. 1986. *Collected Scientific Papers*, Vol. V. Cambridge, MA: MIT Press.

Savage, Leonard. 1954. *Foundations of Statistics*. New York: Wiley.

Schmoller, Gustav von. 1883. "Zur Methodologie der Staats- und Sozial-Wissenschaften" [On the methodology of political and social sciences]. *Jahrbuch für Gesetzgebung, Verwaltung und Volkswirtschaft im Deutschen Reich*, 7, 239–258. https://www.digizeitschriften.de/dms/img/?PID=PPN345575393_0007%7CLOG_0044.

———. 1884. "Comments on Menger's 'Die Irrthümer des Historismus.'" *Jahrbuch für Gesetzgebung, Verwaltung und Volkswirtschaft im deutschen Reich* 8, 677. https://www.digizeitschriften.de/dms/img/?PID=PPN345575393_0008|LOG_0031&physid=PHYS_0694#navi.

Selten, Reinhard. 2001. "What Is Bounded Rationality?" in *Bounded Rationality: The Adaptive Toolbox*, ed. G. Gigerenzer and R. Selten. Cambridge, MA: MIT Press.

Simon, Herbert. 1945. "*Theory of Games and Economic Behavior* by John Von Neumann; Oskar Morgenstern." *American Journal of Sociology* 50 (6): 558–560.

———. 1955. "A Behavioral Model of Rational Choice." *Quarterly Journal of Economics* 69 (1): 99–118.

———. 1978. "Biographical." https://www.nobelprize.org/prizes/economic-sciences/1978/simon/biographical/.

Smith, Adam. 1759. *The Theory of Moral Sentiments*, printed for Andrew Millar, in the Strand; and Alexander Kincaid and J. Bell, in Edinburgh.

———. 1776. *Wealth of Nations*. London: W. Strahan and T. Cadell.

Starr, Ross M. 2008. "Arrow, Kenneth Joseph (born 1921)," in *The New Palgrave Dictionary of Economics* (2nd ed.), eds. Steven N. Durlauf and Lawrence E. Blume. London: Palgrave Macmillan, 232–241, https://econweb.ucsd.edu/~rstarr/ARTICLEwnotes.pdf.

Stone, Richard. 1948. "The Theory of Games." *The Economic Journal* 58 (230): 185–201.

Szpiro, George G. 1986. "Measuring Risk Aversion: An Alternative Approach." *The Review of Economics and Statistics*, 68: 156–159.

———. 1988. "Risk Aversion as a Function of Variance and Skewness." In Atila Chikan, ed. *Progress in Decision, Utility and Risk Theory*. Heidelberg: Springer

———. 2003. *Kepler's Conjecture*. Hoboken, NJ: John Wiley.

———. 2006. *The Secret Life of Numbers: 50 Easy Pieces on How Mathematicians Work and Think*. Washington DC: Joseph Henry Press.
———. 2007. *Poincare's Prize*. New York: Dutton.
———. 2010a. *A Mathematical Medley*. Providence, RI: American Mathematical Society.
———. 2010b. *Numbers Rule: The Vexing Mathematics of Democracy, from Plato to the Present*. Princeton, NJ: Princeton University Press.
———. 2011. *Pricing the Future: Finance, Physics, and the 300-Year Journey to the Black-Scholes Equation*. New York: Basic Books.
———. 2013. "Value Judgments." *Nature* 500, 421–523.
Thaler, Richard. 1980. "Towards a Positive Theory of Consumer Choice." *Journal of Economic Behavior and Organziation* 1 (1): 39–60.
———. 2016. *Misbehaving: The Making of Behavioral Economics*. New York: Norton.
Thaler, Richard H., and Cass R. Sunstein. *Nudge: Improving Decisions About Health, Wealth, and Happiness*. New Haven, CT: Yale University Press, 2008.
Tversky, Amos, and Daniel Kahneman. 1974. "Judgment Under Uncertainty: Heuristics and Biases." *Science*, New Series 185 (4157): 1124–1131.
———. 1981. "The Framing of Decisions and the Psychology of Choice." *Science* 211: 453–458.
United Press International. 1971. "The Pentagon Papers." *1971 Year in Review*, https://www.upi.com/Archives/Audio/Events-of-1971/The-Pentagon-Papers/.
United Nations. 1948. *Universal Declaration of Human Rights*. http://www.un.org/en/udhrbook/pdf/udhr_booklet_en_web.pdf.
UShistory.org. *Declaration of Independence*, 1776, http://www.ushistory.org/declaration/document/.
von Neumann, John. 1928. "Zur Theorie der Gesellschaftsspiele [On the theory of parlor games] *Mathematische Annalen*: 295–320.
von Neumann, John, and Oskar Morgenstern. 1944. *Theory of Games and Economnic Behavior*. Princeton, NJ: Princeton University Press.
———. 1947. *Theory of Games and Economic Behavior* (2nd rev. ed.). Princeton, NJ: Princeton University Press.
Walras, Léon. 1860. "De la cherté des loyers à Paris", *La Presse*, October 19, 1860, https://gallica.bnf.fr/ark:/12148/bpt6k4788476.item. Also front pages October 26, 29, November 6
———. 1874a. "Principe d'une théorie mathématique de l'échange" [Principle of a mathematical theory of exchange]. *Journal des Économistes* 34: 5–22.
———. 1874b. *Éléments d'Économie Politique Pure, ou Théorie de la Richesse Sociale* [Elements of Pure Economics, or the Theory of Social Wealth], 2 vols., Lausanne: L. Corbaz, (2nd ed. Lausanne: L. Corbaz, 1889).

———. 1885. "Un economiste inconnu: Hermann-Henri Gossen." *Journal des Économistes* 30: 68–90.

———. 1896. *Études d'économie sociale.* Paris: F. Rouge, 1896.

———. 1898. *Études d'économie politique appliquée.* Paris: F. Rouge, 1898.

———. 1908. "Un initiateur en economie politique, A. A. Walras." *La revue du mois*, August 10.

Watson, John. 1895. *Hedonistic Theories from Aristippus to Spencer.* London and New York: Macmillan.

Weber, Ernst Heinrich. 1830. *Handbuch der Anatomie des menschlichen Körpers* [Manual of general anatomy of the human body]. Leipzig, Germany: Köhler.

———. 1834. *De subtilitate tactus* [On the precision of the sense of touch].

———. 1846. *Tastsinn und Gemeingefühl.* Reprinted 1905 (ed. Ewald Hering).

Wells, Tom. 2001. *Wild Man: The Life and Times of Daniel Ellsberg.* London: Palgrave Macmillan.

Whately, Richard, Archbishop of Dublin. 1831. *Easy Lessons on Money Matters, for the Use of Young People.* London: Society for Promoting Christian Knowledge.

Wittgenstein, Ludwig. 1921. *Tractatus logico-philosophicus."* (First published in German in 1921 as *Logisch-Philosophische Abhandlung.* First English edition: New York: Harcourt, Brace, and Co., 1922.)

INDEX

absolute risk aversion, 161
Academy of Lausanne, 78–81, 86
accumulation, 31
Adamson, Robert, 106
aggregate happiness, 47
À la Recherche d'une Discipline Économique (In quest of an economic discipline) (Allais), 166
Allais, Maurice: background of, 165–166; education of, 174; independence of irrelevant alternatives challenged by, 171–172; Nobel Prize in Economics of, 167, 224n5; plagiarism controversy of, 224n4; U.S. visit of, 165
Allais effect, 168
Allais Paradox, 203, 227nn19–21
American Economic Association, 211
Analyse, revue et augmenté de plusieurs lettres (Analysis, revised and augmented by several letters) (Montmort), 10
anatomical institute, 54
anchoring heuristic, 196
Andermann, Hermione, 219n22
Annals of Mathematics Studies, 126
Appeal to the Aristocratic Youth (Menger, C., and Rudolf), 96
Aristippus, 23–24, 25, 32, 36, 40
Aristippus the Younger, 26
Aristotle, 41–44, 42, 217n3

Arrow, Kenneth, 155, 171; background of, 154–156; honorary degrees of, 156–157; Nobel Prize in Economics of, 156; risk aversion work of, 158–161; *Schwarzian derivative of a function* and, 224n11
Ashley-Cooper, Anthony, 30, 33
d'Aulnis de Bourouill, Johan, 83, 95
availability heuristic, 194–196
average values, 194
axiomatic systems, xi
axioms, 44–45, 130–135

Baker, Lettice, 111–112
behavioral economics, 19, 92, 97, 207, 214
Behavioral Model of Rational Choice, A (Simon), 179
Behavioural Insight Team, 213–214
belief measures, 114–116
Bentham, Jeremy, 35; Axioms of, 44–45; Declaration of Independence opposed by, 36–37; felicific calculus from, 37–39; *A Fragment on Government* by, 36; *greatest happiness principle* from, 36; happiness quantified by, 37–39; human behaviors noted by, 48; law reforming of, 34–35; physical punishment suffered by, 34; *Principles of the Civil Code* by, 39, 44, 47; psychology of utility from, 206; St. Petersburg Paradox solution of, 44; will instructions of, 36; writings of, 34–35

240   INDEX

Bentham, Samuel, 34
Bernoulli, Daniel, ix–x, *4*, 44, 100; N. Bernoulli contacting, 14–15; books by, 15–16; Cramer's idea accepted by, 15–16; diminishing marginal utility and, 225n11; diversification mentioned by, 21–22; expected value issues of, 16–17; gain compared to loss and, 15–16; human psychology considered by, 169; insurance and, 21; lottery ticket yields from, 17; mean utility from, 17; portfolio theory and, 151; Ramsey and, 115; risk aversion from, 138; St. Petersburg Paradox and, 197, 206; utility of dollar amount and, 68; utility of wealth from, 18–19, 63
Bernoulli, Johann, 15
Bernoulli, Nikolaus, x, 48, 215n7; D. Bernoulli contacted by, 14–15; Cramer's idea striking, 13–14; debt settlement and, 16; *De Usu Artis Conjectandi in Jure* by, 3–4; expected win and, 10; family of, 3; game of chance outcomes of, 4–7; Montmort letter from, 5, 7–8; utility of dollar amount and, 14
biases, 187, 196, 198, 205
Big Short, The (movie), 213
billiard players, 212
Birch, Elizabeth, 33
Black, Fisher, 87, 184
body awareness, 55
Bohr, Niels, 125
Bonaparte, Louis, 77
Book IV (Aristotle), 41
Book VII (Aristotle), 44
bounded rationality, 181, 185, 188, 204–205
Brehm, Alfred, 90
Breitkopf und Härtel (music publishing house), 59, 61, 218n23business cycles, 221n4
Butterfly's Ball and the Grasshopper's Feast, The (Roscoe), 218n1

calculus, 37–39, 74, 87
call options, 184–185

Cameron, David, 213
capital budgeting, 102
cardinal scales, 114, 128, 154
Caritat, Marie Jean Antoine Nicolas de, 131
Carnap, Rudolf, 123
Carnegie Endowment for International Peace, 125
Carnegie Institute of Technology, 181
CCNY. *See* City College of New York
Cech, Eduard, 123–125
Celsius, 134, 153, 222n8, 223n2
certainty, 19
Characteristicks of Men, Manners, Opinions, Times (Locke, J.), 33
Charles II (king of England), 30
Chemin der fer du Nord (railway company), 78
Chladni, Ernst, 52
City College of New York (CCNY), 154
Coal Question: An Inquiry concerning the Progress of the Nation and the Probable Exhaustion of our Coal Mines, The (Jevons), 70
coefficient of utility, 83–84
coenaesthesis, 55
cognitive biases, 205
coin tosses, 199
commodities, 81–82
completeness, 130–131
Condorcet cycle, 131
consumer behaviors, 212
consumers, 81
consumption, 220n5
contentment, 217n3
continuity, as axiom, 132
Contribution to the Theory of Taxation, A (Ramsey), 113
Copeland, Arthur, 137
Copernicus, 106
coping mechanisms, 195
cost-benefit analysis, 102, 208
Cournot, Antoine, 75–76
Cowles Foundation for Research in Economics, 156, 180

Cramer, Gabriel, x, *11*, 153; D. Bernoulli accepting idea of, 15–16; N. Bernoulli and idea of, 13–14; expected win calculations of, 12–13; gain compared to loss and, 15–16; number representations from, 10–12; square root of wealth from, 12–13, 22; utility of dollar amount and, 14, 68
Cuban Missile Crisis, 174
currency, 46, 167
curvature, 223n8

Dameth, Henri, 79
Darwin, Charles, 87, 137
Debreu, Gérard, 167, 225n5
debt settlement, 16
decision-makers, xi; anchoring heuristics of, 196; ecological rationality of, 228n24; gains and losses considered by, 200; heuristics used by, 186, 193–196; human behavior of, 185; Kahneman and Tversky's model for, 199–200; mathematical rules used by, 17; paradigm shift of, 181–182; probabilities combined by, 200; probability theory of, 121; psychological element in, 168–169; in public administration, 180; Simon focusing on, 184; total utility maximized by, 100–101; utility functions and, 146; utility influence on, 108
decision weights, 203–204, *205*
Declaration of Independence, 28, 36–37, 39–40
degree of belief, 114–116, 158, 221n12
Dehaene, Stanislas, 218n25
De la mesure de l'utilité des travaux publics (On measuring the usefulness of public works) (Dupuit), 102
demand curves, slope downward, 103
Denby, Elizabeth, 111–112
depression, 59
Descartes, René, 29
De Subtilitate tactus (Weber, E.), 56
De Usu Artis Conjectandi in Jure (On the use of the technique of conjecturing in matters of law) (Bernoulli, N.), 3–4

diminishing marginal utility, 95, 101, 105; Dupuit's work in, 104; of money, 115; theory of, 84–85, 94, 109; of wealth, 225n11
Diogenes Laertius, 23, 26
Director, Rose, 141
Discourses Concerning Government (Sidney, A.), 30
dispute about methods (Methodenstreit), 97
diversification, 21–22
dolor, 38
Doyle, Arthur Conan, 123
"Dr. Mises" pseudonym, 60
Dupuit, Jules: *De la mesure de l'utilité des travaux publics* by, 102; demand curves from, 103; diminishing marginal utility work of, 104; flood management focus of, 102

Easy Lessons on Money Matters (Whately, R.), 68
École Nationale Superieure des Mines, 166
École Normale, 75
École Polytechnique, 165–166
ecological rationality, 228n24
econometrics, 219n24
economic man, 181, 184
economic mental illusions, 211
economics: analysis of, 136–137; behavioral, 19, 92, 97, 207, 214; humans shaping, 101; laws of, 95, 97; mathematical arguments in, 72–73, 83, 209–210; mathematics in, 87, 182; normative, 184; political philosophy in, 43–44; rational, 210; reasoned opinions in, 96; science of, 71–72, 166–167, 228n2; St. Petersburg Paradox and, 67; theory of, 16, 171, 212; utility of money in, 67; L. Walras devoted to, 77–78
Économie et Interêt (Economy and interest) (Allais, M.), 166
editing process, 200–201, 226n13
educated guesses, 187
Einstein, Albert, 125, 168, 224n4
elections, presidential, 222n13

Elemente der Psychophysik (*Elements of psychophysics*) (Fechner), 61, 63
Éléments d'Économie Politique Pure (*Principles of political economy*) (Walras, L.), 84
Éléments d'Économie Politique Pure, ou Théorie de la Richesse Sociale (*Elements of pure economics, or the theory of social wealth*) (Walras, L.), 80–81
Elizabeth (empress), 89
Ellsberg, Daniel, 172, *173*; car accident influencing, 173–174; felony charges dismissed for, 176; "Risk, Ambiguity, and the Savage Axioms" by, 176; subjective probabilities, 178; Vietnam service of, 175
emotional sensations, 153
endowment effect, 208, 210
Entwickelung der Gesetze des menschlichen Verkehrs, und der daraus fliessenden Regeln für menschliches Handeln (*Development of the laws of human commerce, and of the resulting rules of human action*) (Gossen), 104–106
Epicureanism, 28
Epicurus, 23, 26–28, 32, 40, 181
equilibrium, 123
Essai philosophique sur les probabilités (*Philosophical essay on probability*) (Laplace, S.), 48
Essay d'analyse sur les jeux de hazard (*Essay on the analysis of games of chance*) (Montmort), 7
Études d'économie politique appliquée (Walras, L.), 86
Études d'économie sociale (Walras, L.), 86
Eubotas of Crete, 24
Euler, Leonhard, 3, 15
European currency, 167
evaluation process, 200
exchange rate, 73–74, 82
expected payouts, 171
expected profit, 18
expected utility, 128, 172–176, 200, 203; mathematical side of, 48; of wealth, 18–20

expected utility of the payout, 143
expected value, 16–17
expected wealth, 159, 201
expected win, 9–10, 12–13
experiments, of Weber, E., 56–57, 61–62
exponential function, 215n12
external goods, 44

Fahrenheit, 134, 153, 222n8, 223n2
fair price, 215n5
Fechner, Gustav Theodor, 51, *60*, 127, 202; background of, 58–59; "Dr. Mises" pseudonym of, 60; *Elemente der Psychophysik* by, 61, 63; JND of weight-perception by, 62; mechanical assembly worldview of, 61; psychophysics defined by, 62; *Repertorium der Experimentalphysik* by, 59; sensory stimuli from, 63; university teaching by, 59
felicific calculus, 37–39
Fellow of the American Statistical Association, 157
Fellow to the Royal Society, 29
Ferbach, Célestine-Aline, 77
Fermat, Pierre de, 5–7, 9
Final Problem, The (Doyle, A. C.), 123
financial institutions, 51, 78
financial theory, 16
flood management, 102
fluxional calculus, 74
Foundations of the Theory of Probability (Kolmogorov, A.), 117
Fragment on Government, A (Bentham, J.), 36
fragments, 46
framing, of questions, 198
Francis Sauveur (Walras, L.), 77
Franco-Prussian War, 79
Franz Ferdinand, 91
Franz-Josef (emperor of Austria), 89–91
Frederick III (emperor of Germany), 122, 125
free trade, 86
Freud, Sigmund, 111

Friedman, Milton, 139–140, *140*, 210; insurance-gambling paradox and, 146, 182–183; as monetarism proponent, 141; Nobel Prize in Economics of, 141; utility curve wobble of, *149*; utility function curve of, 146–147
Frisch, Ragnar, 171
Fulbright, William, 175
Fuller, Russell, 213
Fuller & Thaler (asset management company), 213

gamblers: amount willing to pay by, 9–10, 13; debt settlement of, 16; expected value paid by, 16–17; humans as, 138; insurance and, 145–146, 149–150; mental accounting of, 209; odds of wagers, 221n13; probabilities disregarded by, 13–14; utility function and, 145; utility of dollar amount and, 14
gambler's fallacy, 199, 226n12
games of chance, 4; denouncement of, 139; expected win and, 10; future event probability in, 8; mathematical expectation in, 5, 7; paradox, 14–15; possible outcomes of, 6–7; winning chances from, 7
game theory: axioms for, 130–133; book published on, 135–138; editing process in, 226n13; O. Morgenstern on, 125–126; paradox in, 131; von Neumann and, 119, 125
General Mathematical Theory of Political Economy, A (Jevons, S.), 71
generosity, 43
Gigerenzer, Gerd, 228n24
gold coins, 69
Gombeaud, Antoine, (Chevalier de Méré), 5
Gomez, Selena, 213
Gondrecourt, Leopold Graf, 89
goods, 93–94
Gossen, Hermann Heinrich, 102, 104–105, 108
Gossen's First Law, 105
Gossen's Second Law, 105

graveyard of factories, 165
Great Depression, 165
greatest happiness principle, 36
Grundsätze der Volkswirtschaftslehre (Principles of economics) (Menger, C.), 89, 95
guilds, 50

Habsburg dynasty, 91
Halévy, Élie, 38
Handbuch der allgemeinen Anatomie des menschlichen Körpers (Manual of general anatomy of the human body) (Weber, E.), 54
happiness: aggregate and total, 47; J. Bentham quantifying, 37–39; as elusive objective, 40; greatest happiness principle and, 36; from wealth, 45–46
Härtel, Hermann, 61
Hayek, Friedrich, 101
hedonism, 26, 28, 38
heuristics: anchoring, 196; availability, 194–196; decision-makers using, 186, 193–196; humans utilizing, 187, 205; representativeness, 194, 196; as shortcuts, 188
Hicks, John, 156, 167
Hilbert, David, 119
Hochstetter, Christian Gottlieb Ferdinand Ritter von, 90
Hollander, Joseph, 78
Homeyer, Eugen Ferdinand von, 90
homo economicus, 120, 184, 204, 221n2
homo sapiens, 204
honorary degrees, 156–157
Hotelling, Harold, 154
human behaviors, 179; J. Bentham noting, 48; of decision-makers, 185; economic mental illusion and, 211; pleasure and, 36
human nature, 32, 44–45, 201
humans: behavioral concepts of, 36; body, 56; choice shortcuts taken by, 186; each individual owns at least himself and, 31; economics shaped by, 101; gambling

humans (continued)
of, 138; Gossen and interactions of, 104; heuristics utilized by, 187, 205; independence of irrelevant alternatives violated by, 182; interactions of, 105–106; irrational choices of, 183; mental representations made by, 193; numbers concept of, 218n25; pleasure seeking objective of, 24, 28; psychology of, 169; as risk averse, 199; senses of, 55; utility and, 82, 107; utility function of, 134; view of money by, 117–118; weight differences and, 57–58
Hutcheson, Francis, 34
Huygens, Christiaan, 215n5

Imperial Academy of Sciences, 90
Impossibility Theorem, 157
independence, as axiom, 132–133
independence of irrelevant alternatives axiom, 143, 169–170; Allais challenging, 171–172; humans violating, 182; violation of, 177–178, 194
indifference curves, 128
infinitesimal calculus, 87
information, availability of, 194–195
insurance: D. Bernoulli and, 21; gamblers and, 145–146, 149–150; life, 50; premiums for, 138; risk taken on by, 22
insurance-gambling paradox, 146, 182–183
interpersonal comparisons, 153–154
irrational choices, 183

Jahrbuch für Gesetzgebung, Verwaltung und Volkswirthschaft im deutschen Reich (*Annals of legislation, administration and political economy in the German Empire*), 99–100
Jefferson, Thomas, 23, 39
Jeffreys, George, 216n8
Jevons, William Stanley, x, 67, *69*, 106, 220n7; background of, 68; drowning of, 74–75; economic science influenced by, 71; fluxional calculus from, 74; *A General Mathematical Theory of Political Economy* by, 71; mathematical tools and, 73; Mathematico-Economic Writings bibliography by, 95; *The Principles of Science* by, 71; theory of exchange by, 73; *Theory of Political Economy* by, 71, 74, 83–85, 95, 103, 107; utility theory of, 71–72; L. Walras identical theory with, 83–85; writings of, 70
Jevons Paradox, 71
JND. *See* just noticeable difference
John Bates Clark Medal, 156
"Judgment Under Uncertainty: Heuristics and Biases" (Kahneman and Tversky), 193
Justinian I (482–565 CE), 139
just noticeable difference (JND), 57–58, 62

Kahneman, Daniel, 188, *190*, 226n3; background of, 189; decision-making model from, 199–200; decision weights from, 203–204, *205*; human's mental representations from, 193; "Judgment Under Uncertainty: Heuristics and Biases" by, 193; Nobel Prize in Economics of, 192; question framing of, 198; simple theory proposed by, 192–193; value function defined by, 202, *203*
Kapital, Das (Marx), 72
Karl Ludwig, 91
Kautilya 139
Keynes, John Maynard, 70, 111, 113
Knight, Frank, 225n13
Kolmogorov, Andrei, 117–118, 217n12, 221n15
Kortum, Hermann, 107
Kuhn, Harold, 136

Laïs (Aristippus's mistress), 24
Laplace, Pierre-Simon, 48, 49, 67; financial institutions comments of, 51; life insurance interest of, 50; logarithmic utility used by, 217n14; utility theory use by, 49–50
laws, 34–35, 59, 105–106, 116; of economics, 95, 97

League of Nations, 125
Légion d'Honneur, 75–76
Leibniz, Gottfried Wilhelm, x, 3, 87, 113
Lewis, Michael, 225n2, 228n3
liberalism, 30
libertarian paternalism, 213
life insurance, 50
Locke, John, 23, *29*; background of, 28–30; *Characteristicks of Men, Manners, Opinions, Times* by, 33; each individual owns at least himself from, 31; experience gained by, 30; in Fellow to the Royal Society, 29; as medical doctor, 33; *Two Treatises of Government* by, 30, 32–33
logarithmic function, 49
logarithmic utility, 18, 22, 217n14
Long-Range Forecasting Group, 154–155
lotteries, 139, 145
lottery tickets, 17, 145
Luther, Martin, 51

Mailly, Léonide Désirée, 86
majority rule, 131
Manhattan Project, 119
marginalism, 95, 182
marginal utility, 74, 220n7; of commodities, 81–82; of money, 143; theory of diminishing, 84–85; of wealth, 144, 170–171
Markowitz, Harry, 147, 169; Nobel Prize received by, 151, *151*; real-life verification lacking from, 150; utility curve wobble and, 148–150, *149*
Marschak, Jacob, 137
Marx, Karl, 47, 72, 216n1
Mary (princess of Orange), 30
mathematical arguments: in economics, 72–73, 83, 209–210; in risk aversion, 159–161
Mathematical Theory of Saving, A (Ramsey), 113
Mathematico-Economic Writings, 95
mathematics: in economic law, 95; in economics, 87, 182; expectation in, 5, 7; expected utility with, 48; Gossen on importance of, 105; C. Menger and lack of, 94–95; models of, x; rules of, 17; Savage's PhD in, 142; theories in, 80–81; tools in, 73
maximum fee, 12–14
McNamara, Robert, 172, 175
mean utility, 17
measures, 114–116, 153
mechanical assembly, 61
mechanoreceptors (nerve endings), 56
Menger, Anton, 88
Menger, Carl, x, 67, *88*; economic behavior and, 92; Gossen's work reaction of, 108; as journalist, 88–89; mathematical economics from, 87; mathematical sophistication lacking in, 94–95; PhD earned by, 90; polemical style of, 97; prerequisites of goods by, 93; Schmoller disliking, 99; Schmoller refusing book of, 99–100; writings of, 89, 96–97
Menger, Karl, 123, 219n22
Menger, Max, 88
mental accounting, 209–210
mental representations, 193
Merton, Robert, 87
meter, 223n1
meter bar (*Mètre des Archives*), 152
Methodenstreit (dispute about methods), 97, 108
Mètre des Archives (meter bar), 152
military service, 216n21
Minimax Theorem, 121
Mises, Ludwig von, 101, 127
Mohs, Friedrich, 114
Mohs scale, 152, 153
monetarism, 141
monetary income, 226n16
money: diminishing marginal utility of, 115; human's view of, 117–118; marginal utility of, 143; mathematicians estimating quantities of, 10; *numéraire* as, 80; personal consumption and, 31; pleasures from, 28; utility of, 20–21, 67
Montmort, Pierre Rémond de, 3, 5, 7–9

Morgenstern, Oskar, 121, *122*, 153; axioms of, 130–133; axioms violations decree of, 133–134; background of, 123; Cech meeting desired by, 123–125; game theory of, 125–126; independence of irrelevant alternatives axiom of, 143; Jewish rumor of, 127; liberal attitude of, 125; as politically unbearable, 125; *Theory of Games and Economic Behavior* by, 135–138, 156, 168, 222n17; unsavory youth of, 126–127; utility events from, 129; von Neumann with, *124*
Morgenstern, Wilhelm, 122
muscle power, 57

Napoleon III (emperor of France), 77
National Bureau of Economic Research (NBER), 141, 210
National Security Council, 174
natural material, 31
natural rights, 36
Naturphilosophie, 61
NBER. *See* National Bureau of Economic Research
nerve endings (*mechanoreceptors*), 56
nervous breakdown, 59
Newton, Isaac, x, 3; in Fellow to the Royal Society, 29; infinitesimal calculus by, 87; *Principia* by, 137
New York Times, 136–137
Nicomachean Ethics (Aristotle), 41
Nobel, Alfred, x, 86
Nobel Peace Prize, 86
Nobel Prize, xi, 86; H. Markowitz receiving, 151, *151*; Selten winning, 184
Nobel Prize in Economics, 182, 223n3; of Allais, 167, 224n5; of Arrow, 156; of Friedman, 141; of Kahneman, 192; of Simon, 179; of Thaler, 211, 228n3
normative economics, 184
nuclear weapons, 174
numbers, representation of, 10–12
numéraire (money), 80

Obama, Barack, 213–214
Ohm's law, 59

Oken, Lorenz, 61
"On the Optimal Use of Winds for Flight Planning" (Arrow, K.), 156
On the Origin of Species (Darwin, C.), 137
opportunity cost, 208, 210
ordinal scale, 114
österreichische Adel und sein constitutioneller Beruf: Mahnruf an die aristokratische Jugend, Der (*The Austrian nobility and its constitutional mission: Appeal to the aristocratic youth*) (pamphlet), 91
outcomes, possible, 6–7
overestimation, 227n22
overweighting, 227n22

pain, 24, 36
Pannomial Fragments (Bentham, J.), 44, 46–47
paraconical pendulum, 168
paradox: Allais, 203, 227nn19–21; expected utility influenced by, 172–176; games of chance, 14–15; in game theory, 131; insurance-gambling, 146, 182–183; Jevons, 71; Super-Petersburg, 219n22. *See also* St. Petersburg Paradox
Pareto, Vilfredo, 86
parlor games, 120–121
Pascal, Blaise, 5–7, 9
payoffs, minimal, 121
pensions, old-age, 50
Pentagon Papers, 172–176
personal consumption, 31
philanthropy, 42–43
physical punishment, 34
plagiarism, 224n4
pleasure, 126; hedonism with, 26; human behavioral concept of, 36; human's objective of, 24, 28; from money, 28; sober reasoning and, 26
Poincaré, Henri, 168
Poisson, Siméon-Denis, 67
political philosophy, 43–44
Politika (Aristotle), 43
Pont de Nemours, Pierre Samuel du, 40
Popper, Karl, 123
portfolio theory, 151

possible outcomes, 6–7
Pratt, John Winsor, 154, 157–161
predictions, 123, 146, 196
premiums, for insurance, 138
presidential election, 222n13
prices: of call options, 184–185; fair, 215n5; supply and demand influencing, 80; traders getting correct, 80
Principia (Newton), 137
Principles of Morals and Legislation, 38
Principles of Science, The (Jevons, S.), 71
Principles of the Civil Code (Bentham, J.), 39, 44, 47
probabilities, 38; decision-makers combining, 200; decision weights and, 203; degree of belief in, 114–116, 158, 221n12; diversification influencing, 21–22; expected value of, 16–17; expected win, 9; gamblers disregarding, 13–14; games of chance and, 8; Kolmogorov's theory on, 117–118; laws of, 116; measuring, 114; St. Petersburg Paradox and, 226n14; subjective, 177–178
probability theory, 8, 121, 199, 217n12, 221n15
progressive taxation, 46
property, preservation of, 31–33
property rights, 32
propinquity, 38
proportions, 58
prospect theory, 200
"Prospect Theory: An Analysis of Decision Under Risk" (paper), 200
psychoanalysis, 111, 174
psychological element, 168–169
psychophysics, 62
public administration, 180
Pyke, Margaret, 110–111

quantum mechanics, 125

Ramsey, Frank, *110*; background of, 109–111; belief measures by, 114–116; D. Bernoulli and, 115; death of, 112; human's view of money by, 117–118; Keynes as mentor to, 113;

Pyke as married woman and, 110–111; subjective probabilities from, 177; writings by, 113
Rand, Ayn. *See* Rosenbaum, Alisa Zinoyevna
RAND Corporation, 156, 172, 175
rareté (scarcity), 81, 84
rational economics, 210
rationality, 183–184, 186
rational maximizing model, 211
Rau, Karl Heinrich, 220n5
raw materials, 82
reasoned opinions, 96
Recherches sur les principes mathématiques de la thorie des richesses (*Research on the mathematical principles on the theory of wealth*) (Cournot), 77, 80
redistribution, of wealth, 47
regression to the mean, 197, 199
Reik, Theodor, 111
relative hardness, 152
relative risk aversion, 161
Repertorium der Experimentalphysik (Fechner, G.), 59
representativeness heuristic, 194, 196
Ricardo, David, 67
risk, 22, 225n13
Risk, Ambiguity, and Decision (thesis), 174
"Risk, Ambiguity, and the Savage Axioms" (Ellsberg, D.), 176
risk aversion, 154, 224n9; absolute, 161; Arrow's work on, 158–161; from Bernoulli, D., 138; formula for, 160; humans as, 199; lotteries and, 145; mathematical arguments in, 159–161; Pratt's work on, 158–159; utility and, 19–21; utility function curve and, 160–161
"Risk Aversion in the Small and in the Large" (Pratt), 157–158
risky behaviors, 146
Roosevelt, Teddy, 86
Roscoe, William, 68, 218n1
Rosenbaum, Alisa Zinoyevna, 101
Rothschild, Nathaniel Meyer, 91
Ruchonnet, Louis, 78–79

Rudolf (crown prince of Austria-Hungary), 89–91
Russell, Bertrand, 110, 112

Samuelson, Paul, 167, 171
satiation, 44, 220n5
satisficing, 181, 186, 188, 204
Savage, Leonard ("Jimmie"), 139, *142*, 171; background of, 141–143; insurance-gambling paradox and, 182–183; mathematics PhD of, 142; utility function curve of, 146–147; as von Neumann assistant, 143
scarcity (*rareté*), 81, 84
Schmidt, Friederike, 53
Schmoller, Gustav von, 96–97, 99–100, 108
Scholes, Myron, 87, 184
Schwarzian derivative of a function, 224n11
Schweitzer, Selma, 156
scientific methodology, 98
scientific precursors, 107
Selten, Reinhard, 184
sensory experiences, 96
sensory organs, 55
sensory stimuli, 63
Serious Fall in the Value of Gold ascertained, and its Social effects set forth, with two Diagrams, A (Jevons, S.), 70
Shaftesbury, earl of, 33–34
shortcuts, 186, 188, 195
Sidney, Algernon, 30
Simon, Herbert, 137, *180*, 204, 214; background of, 179–181; *A Behavioral Model of Rational Choice* by, 179; at Carnegie Institute of Technology, 181; concepts of, 181; decision-making focus of, 184; Nobel Prize in Economics of, 179; real world occurrences, 207
simplification, 201
Smith, Adam, x, 67, 206
Smith, Vernon, 226n3, 228n24
sober reasoning, 26
social justice, 78
social sciences, 179
Société du Mont Pélerin think tank, 167

sociétés de secours mutuels (*societies for mutual support*), 50
Socrates, 23–24
Solow, Robert, 167
"Some Results in the Decision Theory of One-Parameter Multivariate Polya-Type Distribution" (Arrow, K.), 157
Sophocles (Greek poet), 138
sources of contentment, 217n3
Spann, Othmar, 127
Specimen theoriae novae de mensura sortis (*Exposition of a new theory on the measurement of risk*) (Bernoulli, D.), 16
Specimen theoriae novae metiendi sortem pecunariam (*Exposition of a new theory of measuring financial outcomes*) (Bernoulli, D.), 15
square root of wealth, 12–13, 22
Stein, Gertrude, 11
sterility of exaggerated luxury, 78
stimulation response, 56, 62–63
Stone, Richard, 137
St. Petersburg Paradox, 28, 121, 182–183, 199; J. Bentham solution to, 44; D. Bernoulli and, 197, 206; economics and, 67; mathematicians problem with, 22; probability and, 226n14
subadditivity, 204, 227n20
subcertainty, 203
subjective probabilities, 177–178
subproportionality, 204
sunk cost, 208, 210
sunk cost fallacy, 197, 210
Sunstein, Cass, 212–213
Super-Petersburg Paradox, 219n22
supply and demand, 74, 80
Supreme Court, 175
swordmaker, 218n19
Szeps, Moritz, 91

tactile senses, 56
Tastsinn und das Gemeingefühl, Der (Weber, E.), 55
taxation, 46, 78
Teichler, Margarete, 122

temperature, 129, 218n20, 222n8, 223n2
Thaler, Richard, xi, *209*, 213; background of, 207–208; behavioral economics from, 207, 214; consumer behaviors and, 212; cost-benefit analysis from, 208; Nobel Prize in Economics of, 211, 228n3
Théorie analytique des probabilités (Analytical theory of probability) (Laplace), 48, 50
theory: of bounded rationality, 186; of diminishing marginal utility, 84–85, 94, 109; of economics, 16, 171, 212; of exchange, 73, 82; of expected utility, 169–172; of general equilibrium, 80; of heat, 129; of probabilities, 8, 121, 199, 217n12, 221n15. *See also specific theory*
Theory of Games and Economic Behavior (Morgenstern, C., and von Neumann), 135–138, 143, 156, 168, 222n17
Theory of Moral Sentiments, The, 206
Theory of Political Economy (Jevons), 71, 74, 83–85, 95, 103, 107
Theory of Value (Debreu), 225n5
Thurmburg, Josef Latour von, 89
total happiness, 47
total utility, 100–101, 220n7
"Towards a Positive Theory of Consumer Choice" (paper), 210
Tractatus logico-philosophicus (Wittgenstein), 109
traders, 73–74, 80, 184
transitivity, as axiom, 131–132
Treatise on Probability (Keynes), 113
Truth and Probability (Ramsey), 113
Tversky, Amos, 188, *191*; background of, 189–191; decision-making model from, 199–200; decision weights from, 203, *205*; human's mental representations from, 193; "Judgment Under Uncertainty: Heuristics and Biases" by, 193; question framing of, 198; simple theory proposed by, 192–193; value function defined by, 202, *203*
Two Treatises of Government (Locke, J.), 30, 32–33

UCL. *See* University College London
uncertainty, 21, 225n13
Undoing Project, The (Lewis, M.), 228n3
Universal Declaration of Human Rights, 40
University College London (UCL), 36, 69–70
University of Dresden, 55
University of Lausanne, 85–86
University of Leipzig, 58
Untersuchungen über die Methode der Socialwissenschaften (Inquiries into the methods of the social sciences) (Menger, C.), 97
"U.S. Decision-making in Vietnam, 1945–68" (McNamara), 175
utilitarianism, 34, 36
utility, 47; J. Bentham psychology of, 206; coefficient of, 83–84; curve, 19–20; decision-makers influenced by, 108; diminishing rate of, 48; expected, 48, 203; gamblers and dollar amount, 14; goods must provide, 93; humans and, 82, 107; human's diminishing, 107; indifference curves and, 128; interpersonal comparisons of, 153–154; Jevons theory of, 71–72; measures resistance of, 153; of monetary income, 226n16; of money, 20–21, 67; risk aversion and, 19–21; total, 100–101, 220n7; traders negotiation of, 73–74; value of item based in, 17; von Neumann and events of, 129, 153; of wealth, 17–19, 20, 44–45, 63, *160*; wealth and dollar's, 12–13
utility curve, 148–150, *149*, 197–198
utility function, 128, 223n8, 224n9; axiom adherence for, 133–135; decision-makers and, 146; Friedman and Savage curve of, 146–147; gamblers and, 145; of humans, 134; risk aversion and curve of, 160–161; of wealth, *145*, 146–147; wiggly, 145–146
utility of the expected payout, 143–144
"Utility of Wealth" (paper), 150
utility theory, 49–50, 178

vaccinations, 195
value curve, 202
value function, 202, *203*
value of items, 17
Vetsera, Mary (baroness), 90
Virginia Declaration of Rights, 39
virtuous man, 41
vNM-irrational, 133
vNM-rational, 133
volatility, 185
von Neumann, John, *120*, 123; axioms of, 130–133; axioms violations decree of, 133–134; game theory and, 119, 125; independence of irrelevant alternatives axiom of, 143; minimal payoffs and, 121; Savage as assistant to, 143; *Theory of Games and Economic Behavior* by, 135–138, 156, 168, 222n17; utility events from, 129, 153; von Neumann with, *124*

Wakefield, Andrew, 195
Wallace, Alfred Russell, 87
Walras, Auguste, 75–76
Walras, Léon, x, 67, *76*, 182; background of, 77; celebration for, 87; economics devotion of, 77–78; Jevons's identical theory with, 83–85; as journalist, 77; Kortum tribute by, 107; mathematical theory of economics by, 80–81; scientific precursors of, 107; theory of general equilibrium from, 80; at University of Lausanne, 78–79, 85–86; writings of, 80–81, 84, 86
Walras, Marie-Aline, 86
water usage, 102
wealth: diminishing marginal utility of, 225n11; dollar's utility in, 12–13; expected, 159, 201; expected utility of, 18–20; generosity of, 43; happiness from, 45–46; human nature and increase of, 32; losses larger than gains in, 202; marginal utility of, 144, 170–171; perceptions in differences of, 58; redistribution of, 47; square root of, 12–13, 22; utility curve of, 197–198; utility function of, *145*, 146–147; utility of, 17–19, *20*, 44–45, 63, *160*
Wealth of Nations (Smith, A.), 206
Weber, Ernst Heinrich, 51–53, *52*, 127, 202, 218n20; anatomical institute's condition from, 54; experiments of, 56–57, 61–62; human senses interest of, 55; muscle power findings of, 57; writings of, 54–56
Weber, Michael, 51
weight differences, 57–58
weight-perception, 62–63
Wellenlehre auf Experimente gegründet (Textbook on waves, based on experiments) (Chladni, E. and Chladni, W.), 53
Whately, Richard, 68
Wieser, Friedrich von, 101
wiggly utility function, 145–146
winning chances, 7
Wittenberg, 52–53
Wittgenstein, Ludwig, 109, 112, 123
Wolfensgrün, Menger Edler von, 87

zero-sum games, 121
Zur Methodologie der Staats- und Sozial-Wissenschaften (On the methodology of political and social sciences) (Schmoller), 98
Zur Theorie der Gesellschaftsspiele (On the theory of parlor games) (von Neumann), 120